中国北方地下水可持续管理

Sustainable Groundwater Management for North China

主 编 曲士松

副主编 李庆国　王维平　马振民

编 委 (以姓名拼音顺序)

姜　巍　梁　伟　吕　华

王晓军　解伏菊　杨丽原

于晓燕

黄河水利出版社

内 容 提 要

本书精选了联合国教科文组织第二届"中国北方地下水可持续管理"培训班中 30 篇文章,反映出目前地下水可持续管理的最新研究成果。其主要内容包括地下水可持续发展的原则、地下水和地表水联合运用、地下水污染和保护、地下水保护监测和模型、地下水非点源污染控制、水生态、地下水有效管理的经济效益、城市雨水管理、污水处理及回用、济南供水保泉等。

本书适合从事地下水科研、教学、管理及有关部门人员参考。

图书在版编目(CIP)数据

中国北方地下水可持续管理/曲士松主编. —郑州:黄河水利出版社,2008.6
 ISBN 978-7-80734-413-1

Ⅰ.中… Ⅱ.曲… Ⅲ.地下水资源-水资源管理-研究-中国 Ⅳ.P641.8 TV213.4

中国版本图书馆 CIP 数据核字(2008)第 084128 号

策划组稿:王路平 ☎ 0371-66022212 E-mail:hhslwlp@126.com

出 版 社:黄河水利出版社
    地址:河南省郑州市金水路 11 号 邮政编码:450003
发行单位:黄河水利出版社
    发行部电话:0371-66026940、66020550、66028024、66022620(传真)
    E-mail:hhslcbs@126.com
承印单位:黄河水利委员会印刷厂
开本:890 mm×1 240 mm 1/16
印张:10.50
字数:300 千字 印数:1—1 000
版次:2008 年 6 月第 1 版 印次:2008 年 6 月第 1 次印刷

定价:30.00 元

序 一

  水资源是人类生存和发展不可替代的宝贵的自然资源，又是生态环境的重要组成部分。我国要实现社会经济的可持续发展，无论是经济建设还是生态建设都需要加强对水资源的保护和合理利用。我国淡水资源总量约为 28 124 亿 m^3/a，人均占有淡水资源量为 2 163 m^3/a，不到全球人均占有淡水资源量的 1/4。全国地下淡水天然补给资源量约为 8 840 亿 m^3/a，地下淡水可开采资源量为 3 527 亿 m^3/a，其中北方地下淡水可采资源量为 1 536 亿 m^3/a，占全国淡水可采资源量的 43.5%。目前全国平均地下水开采程度为 36%，北方 15 个省、区、市地下水开采程度则达到 60%，其中华北地区平均高达 76%，东北地区 65%，西北地区 25%，有 37 个城市和地区处于超采状态。因此，中国的水资源问题主要在北方。一方面，自然区域条件形成北方水少，水资源短缺问题突出；另一方面，人为因素又加剧了北方的缺水状况。

  地下水是我国北方整个水资源系统中极其重要的组成部分，随着人口增长和经济发展，地下水开发不合理、供需紧张等问题的加剧，地下水资源的保护和可持续利用越来越受到人们的关注。近年来，中国北方地区超采地下水诱发的地面沉降、地面塌陷、海水入侵等一系列地质灾害现象，给我国的经济、社会和自然环境带来重大损失。因此，加强地下水资源保护和科学利用极其重要。联合国教科文组织于 2007 年 3 月在济南大学召开第二届"中国北方地下水可持续管理"培训会议，中外专家、学者共同研讨地下水可持续发展的原则、地下水和地表水联合运用、地下水污染和保护、地下水保护监测和模型、雨水管理的效果、污水处理及回用、地下水非点源污染控制、水生态、地下水有效管理的经济效益及济南供水保泉等议题，并组织编撰论文集。希望《中国北方地下水可持续管理》一书的出版，能有助于进一步推动地下水资源可持续利用与管理方面的工作，为水资源科学发展做出积极贡献。特为之序。

济南大学校长：

2008 年 2 月，济南

序 二

水是生命之本，是人类的文明之源。地下水作为地球水文循环中的重要一环，在人类赖以生存的水资源中占据着极其重要的地位。在经济社会持续快速发展的今天，人类对水的需求与日俱增，中国正面临着水资源严重短缺的危机。与地表水相比较，由于地下水具有调蓄能力强、供水保证率高、水质优、分布广、开采方便等特点，它在水资源的配置和环境保护中发挥着特殊的作用，它是重要的经济资源、环境资源和战略资源。

山东是一个农业大省、经济大省和人口大省，也是一个水资源严重短缺的省份，山东以仅占全国水资源总量的1.1%的水，养育着占全国7.1%的人口，灌溉着占全国7.4%的耕地，生产着占全国10%的粮食，支撑着全国9.3%的国内生产总值，其中，地下水供水量占总供水量的50%以上，地下水在支撑经济社会持续快速发展方面功不可没。但是也应当看到，多年来由于对地下水资源不合理开采，引发了区域内局部地下水位下降、海水入侵、水质污染等问题。因此，科学规划、合理开发和有效保护地下水资源已经成为近年来水资源管理工作的重点。尽管我们在地下水资源开发利用和节约保护方面做了许多工作，但是离实现人水和谐共处的目标还有差距，如何更好地加强地下水资源开发利用管理和保护仍然是摆在我们面前的一个重大课题。

联合国教科文组织(UNESCO)将第二届"中国北方地下水可持续管理"培训班安排在山东济南举办，组织国际和国内地下水专家有针对性地就地下水可持续利用原则、地下水和地表水联合运用、地下水污染和非点源污染控制以及水生态保护、地下水管理经济费用分析等方面内容进行专题授课，不仅为我省水资源管理工作提供了一次难得的学习机会，而且必将对山东地下水资源开发利用、科学管理和有效保护工作起到极大的促进作用。专家授课讲义的编著成册，也必将为水资源管理工作者提供一本良好的教科书。

借此机会，真诚地感谢联合国教科文组织(UNESCO)对山东水资源管理工作的帮助，感谢专家教授和会议组织者为此付出的辛勤劳动，感谢社会各界的关心和支持。

山东省水利厅副厅长：

2008年2月，济南

前 言

地下水资源是人类饮用水的优质水源，是中国北方地区城市生活和工业主要的供水水源。中国北方地区水资源短缺，由于地下水超采严重，引发了大范围的地下水漏斗、地下水资源枯竭、水质恶化、海水入侵、地面沉降、岩溶塌陷以及生态环境恶化等现象。同时，由于工业化和城市化进程的加快，城市生活污水和工业废水部分未经处理或经处理后污染物总量不能达到自然界本身净化的能力，河流污染致使河流附近呈线状地下水污染；部分企业在生产过程中，污染物的跑、冒、滴、漏以及突发事故致使呈点状地下水污染；农业施用农药、化肥造成的面状地下水污染。地下水过量开采和污染正在威胁着我们的生存环境和健康。特别是这种看不见的地下水污染，越来越引起国内外科学技术界的重视。

联合国教科文组织(UNESCO)于 2005 年设置了水资源可持续管理教席，由 W.F.Geiger 教授担任，目的是增进交流和以大学、企业合作的形式，通过举办一系列培训班、研讨会、展览等方式加强可持续用水管理方面知识的学习。在济南大学举办的"中国北方地下水可持续管理"培训班是该教席的第二次活动。现今世界上有 11 亿人口没有获得安全卫生的饮用水，按联合国千年发展计划，到 2015 年将有 5.5 亿人口的饮用水安全问题得以解决。应重点考虑农业生产率高的地区，这些地方已经对水资源，特别是地下水带来压力。为了解决当前水和未来水的问题，必须建立一个可持续用水管理的，特别是应用到地下水管理方面的新范例。尽管我们的水管理已有很长的历史，然而要想创立一新的范例就必须要有水管理的新方法才行。

由联合国教科文组织(UNESCO)和济南大学举办的第二届"中国北方地下水可持续管理"培训班，邀请了德国、澳大利亚以及国内高校、科研、管理等部门的教授和专家，从不同的学术观点、不同的方面，多角度地对地下水可持续管理的有关问题进行讲座和研讨，涉及到浅层和深层地下水、孔隙水和岩溶水、地表水与地下水联合调度、地下水污染与控制、再生水灌溉对地下水环境的影响、城市雨水回灌岩溶地下水、地下水监测、地下水可持续管理、地下水模型、水生态修复以及新方法、新技术在地下水研究方面的应用等，在研究内容上互相补充，各具特色。

该培训班得到了联合国教科文组织(UNESCO)、济南大学、山东省水利厅、济南市水利局、山东水利学会、山东侨务办公室以及武汉中地公司的支持。本书出版得到了山东省重点学科(试验室)基金项目资助。

书中不当之处，敬请读者批评指正。

作 者

2008 年 3 月

目 录

序一 .... 程　新
序二 .... 刘勇毅
前言

Sustainability Principles for Water Management .... W. F. Geiger (1)
Methodology and Practical Examples of Model-based Groundwater Monitoring, Management,
　　Protection and Remediation .... Stefan Kaden (14)
Impact of Stormwater Recharge on Blue Lake, Mount Gambier's Drinking Water Supply
　　..................................... Joanne Vanderzalm　Tara Schiller　Peter Dillon　Stewart Burn (22)
Coupled Modeling of Groundwater and Surface Water for Integrated, Sustainable Water
　　Management in Wetlands .... B. Monninkhoff　S. O. Kaden (28)
Berlin Water Supply—an Example of Conjunctive Use of Groundwater and Surface Water
　　... Luo Junfeng (34)
Nitrogen Transport in Soils Under the Condition of Sewage Irrigation
　　... Yang Jinzhong　Wang Liying (38)
Seawater Intrusion and Land Subsidence Caused by Groundwater Overpumping in China
　　... Wu Jichun　Xue Yuqun　Shi Xiaoqing (51)
污染场地健康风险评价的理论和方法 .... 陈鸿汉　谌宏伟　何江涛　等 (58)
华北平原地下水超采现状及对策 .... 马凤山 (63)
An Integrated Groundwater Management GIS to Improve Water Supply Safeguard for
　　Emergency Well-field, Beijing .... Wei Jiahua　Li Yu (66)
济南市玉绣河水生态环境修复综合措施 .... 王　琳 (71)
济南市水资源可持续利用实践与探索 .... 孟庆斌 (77)
山东省黄泛平原深层地下水资源可持续利用 .... 徐军祥 (81)
河流生态修复——时代赋予水利的重要使命 .... 宫崇楠 (86)
地下水脆弱性评价方法及其应用 .... 张保祥 (89)
Assessing the State of Water Resources and Sustainable Water Management Strategies in Taicang
　　City, Jiangsu Province .... Lu Zhibo　Wang Juan　Deng Dehan　Liu Ning　Yang Jian (93)
Control of the Water Resources Risk .... Tian jinghong (98)
The Application of Geographic Information System in Urban Water-supplying and Water-draining
　　Industry .... Zeng Wen　Wang Xiguang (104)
岩溶地下水污染控制技术 .... 马振民　于玮玮 (110)
城市雨洪水利用与回补岩溶地下水 .... 王维平　曲士松　邢立亭　等 (114)
济南保泉对策研究 .... 邢立亭 (119)
地下水地表水库联合调度分析 .... 刘本华 (124)
济南市地下水保护供水保泉成本效益分析 .... 李庆国 (130)
济南市地下水补给水源涵养能力分析 .... 周保华　李大秋 (135)
浅议海侵区农民用水者协会在水生态修复中的作用 .... 曲士松　王维平　孙小滨 (139)

· 1 ·

区域农业水资源联合调控技术研究 ………………………………… 徐征和　贠汝安 (143)
降水和开采对济南市区泉群流量的影响及其贡献 ………………… 王晓军　陈学群　张维英 (148)
城市湿地生态系统的生态功能与保护对策 ………………… 王　惠　曲士松　杨宝山　等 (150)
城市化对湿地的影响及保护措施 ………………………………………………… 张明亮 (153)
某饮用水源水库铁、锰垂直分布规律及原因探讨 ……………………………… 王海霞 (155)

Sustainability Principles for Water Management

W. F. Geiger

(UNESCO Chair in Sustainable Water Management)

1 Introduction

Population together with the expansion of irrigated agriculture and industrial development have increased in the past 50 years significantly. While in 1950 30% of the world's population lived in urban areas, today it is more than 50%. This created tremendous pressure on water resources. Worldwide water demand amplified by a factor of six between 1900 and 1995 while population only doubled. Water resources are exploited and polluted, overlooking environmental degradation. Results are flooding, droughts, abuse of water resources, sinking groundwater levels, saltwater intrusion along costs, land subsidence, pollution, which makes water unusable, and global changes, that lead to more frequent occurrences of disasters. The higher economic growth is, the higher environmental damage seems to be.

Despite of overexploitation of water resources inadequate water supply and sanitation have been identified by the World Summit in Johannesburg 2002 to be the major problems. Today, there are 1.1×10^9 people without access to safe water supply, 2.4×10^9 people without adequate sanitation and even more who are short of water for food production. To meet the Millennium Development goals set by the World Summit in Johannesburg, namely to half the number of people lacking access to safe water supply and adequate sanitation by 2015 requires politicians, planners and users to review inherited thinking, planning measures used in the past, which evidently were not suitable or too inflexible to cope with the problem so far.

Also in China in the past 50 years, water demand increased as rapid as economic development and population. In 2004, the total national water consumption has increased to 554.8 m^3 as against 103.1×10^9 m^3 in 1949. Allocation among diverse users changed too. Agriculture still is the major user with 64.6% of the total water consumption.

At present, two thirds of Chinese cities face serious water shortage. Northern China and parts of Western China are known to have serious water shortages. Losses in water storage capacity add to the problem of water shortage. The total storage capacity of 601 large-sized and medium-sized reservoirs in the Yellow River basin i.e. was 52.3×10^9 m^3, 10.9×10^9 m^3 of which (21% of the total) had been lost due to sedimentation until 1989. Regulation of upstream reservoirs made peak floods decrease greatly in the Lower Yellow River during the past two decades, leading most sediment to deposit in the main channel of the river. This and irrational use of floodplains along the river, caused shrinking of the channels. The loss of storage capacities not only aggravates the risk of flooding and droughts, but also ecological deterioration.

Presently, in China an area of 3.56×10^6 km^2 is eroded, accounting for 37% of the total territory, causing annually 5×10^9 t of eroded soil. This led to: an annual loss of arable land of more than 66×10^3 hm^2 in the past 50 years, degradation of grassland of 1×10^6 km^2 in total, land desertification of 2.46×10^3 km^2 a year since 1990, and sedimentation in reservoirs and rivers, reducing their flood regulation and conveyance capacity. The ecological degradation in eroded areas worsens poverty. Over 90% of the poor in China live in such areas.

Pollution is an even more serious problem in China. River reaches of about 1×10^5 km have grade IV or worse accounting for 47% of the total length. More than 75% of the lakes are heavily polluted. A study on drinking water of 118 cities indicates that groundwater has been polluted to a varying degree in 97% of the cities, 64% of which are seriously aggravating water shortages further. In addition water pollution increases ecological deterioration reducing the value of water not only for domestic, industrial and agricultural use, but also for recreation.

2 Imbalances and barriers obstructing water management

Introduced problems lead to social, environmental and economic inequity. While UNESCO in the 70ies of 20th century through its International Hydrologic Programmes (IHP) pointed out these dimensions of water problems and endorsed sustainable and integrated water management and though sustainability was addressed in World Summits and World Water Congresses for decades, problems became larger. While governments passed laws requiring integrated water management and sustainability, and while 179 nations signed the Agenda 21 in 1992 including the paradigm of 'Sustainable Development' as the basis for future policies, reality in acting sustainable despite of all verbal insight and available techniques lacks behind.

2.1 Changing hydrologic and social situation distressing water managers

Above problems mainly are man-made, i.e. land use alterations resulting in more runoff and pollution, population growth and migration from rural to urban areas resulting in more urbanization, illegal discharges of solid and liquid wastes and last not least human behavior and consumerism resulting in global warming affecting sea levels and causing more frequent disasters. All along an ecological crisis and social changes are observed. It is not understood, what causes such socio-ecological transformations, where they lead to and which development potential is contained in these processes. However they are strongly linked to the water crises.

In the 1,034 km^2 Huangshui river basin in Shandong for example per year 120×10^6 m^3 of runoff can be utilized, while water demand with 162×10^6 m³/a exceeds renewable water resources by about 25% (Geiger et al., 2005). In 1998 in Shandong more than 4 million people in 30 cities suffered water scarcity. Citizens had to use water just part time and limited. Groundwater was overused causing saltwater intrusion, which affects agricultural production badly. Food production was reduced by 9.65×10^6 t and annual economic losses only due to water scarcity are estimated over 5 billion Yuan for the province. Even more because lack of water, many have to drink low-quality water. This again increases the incidence of diseases.

Due to economic development and population increase a lot of water originally available for agriculture is now being used for supply of cities and industry. This causes conflicts between towns and counties, production and living, not even considering environmental needs. It is unclear how much options the different social parties have. Action without clear goals and insufficient means lead to progressing dynamics of problems which get out of control. Social justice gets lost and social solidarity decreases. This is the case not only at the Huangshui river.

2.2 Increasing insight and new technologies upsetting decision makers

Water management today can benefit from the rapid development of monitoring, i.e. remote sensing, data processing, i.e. GIS systems and modelling techniques. Further, background knowledge on hydrologic and bio-chemical processes has never been as large as it is today. These advantages, however, are counteracted by some managers and administrators not keeping pace with the necessary knowledge needed to operate latest techniques. In effect decision makers often relay on modelling results produced by technically qualified, but in water management less experienced (computer) specialists. It is merely impossible for them to check if options are reasonable or not. The risk, that money is wasted, increases.

Even more, the definition of the catchments' boundary always was a difficult task, as it depends on the problem to be solved. Management strategies on different scales at last must go with each other. Aside the technical difficulties for integrated planning different responsibilities, political priorities and ignored law offences make it impossible to satisfy all including environment's interest.

Traditional water resources management also suffers from the difference between the long times (up to ten years) that are required to assess existing conditions by monitoring and modelling quantity and quality processes in river catchments and the short times available for deciding on individual measures (sometimes less than one year, in case of chemical accidents minutes). In consequence measures taken cannot be checked if they fit into long-term water management objectives. This usually results in low efficiencies and high costs.

2.3 Administrative constraints hampering water management

Sometimes problems seem to be technical at a first glance, but actually are administrative:

●Data established in earlier plans are lost in many instances. New projects, often dealing with issues that have been addressed previously, have to start from zero.

●Another deficiency is the lack and inappropriate use of data. The time and space resolution needed often are not available from standard data sources.

●Fast growth and changes make it difficult to make correct forecasts. While remote sensing data could always be updated, this often is found too expensive for water management.

●Planning often not adapts to specific conditions. A widespread fault is copying procedures from somewhere else, which usually are inappropriate to solve the problem at hand.

●Financial and social data are equally important. Different dimensions, conventions and responsibilities for these sectors make integrated planning very difficult.

The major constraints are not technical but organizational. Insufficient institutional competence at various government levels not only hinders to develop alternative and locally adapted innovations, but also hamper the creation of adequate traditional infrastructures. The artificial subdivision of reality into different sectors opposes interdisciplinary thinking.

Maybe the biggest administrative constraint is shifting responsibilities, a phenomenon met in developing and developed countries alike (UNESCO, 1996). It implies that when a problem is recognized, parties affected decide that they do not want to be mixed up, especially when high costs may be met or difficult political hurdles are seen. The general attitude in such cases is either to deny responsibility for the recognized problem or to take up the standpoint that the problem is not as big as thought. What it boils down to is not only the question of who is legally responsible, but also who is prepared to take it up and actually starts improvements.

2.4 Political priorities and corruption paralyzing water management

Political constraints often are more significant than technical and administrative altogether:

●Financial and technical tasks often differ. When selecting a solution for a water s/
pollution or remediation problem, a purely technical approach unconnected with financial rea/

create unrealistic solutions. The final decision lies with political decision makers, who may have little or no technical and economic background.

• General organization and management structures often are unable to recognize the relations of technical, financial and social problems and to find integrated solutions. Decision makers are confused by the fundamental differences of engineers and environmentalists about the way in which a recognized problem should be solved.

• There is a tendency to deal ineffectively with risks. Unknown features of the planning process paralyse decision makers, limit innovations and result in stagnant technology and wrong solutions. Risk assessment and public discussion of the risks involved are needed.

• The political pressure, under which planning processes must progress, often blocks efficient planning.

• In general the planning process tends to be reactive rather than proactive. This unfortunately is observed the more, the higher the responsibility level is. Because of the obstacles listed above, science, technology, engineering design, ecological analysis, impact assessment and other factors all operate as reactive systems.

More severe is the ignorance of water laws and environmental protection requirements set out by government. In countries, where economic development receives the highest priority, corruption is observed in context with unpunished law offences. Frequent spills, accidents and related social and environmental disasters are difficult to be explained otherwise.

3 Principles and objectives for sustainable water management

Today sustainability i.e. the balance linking environmental, social and economic needs almost never is achieved, also because the specific characteristics of individual water systems make it impossible to have a universal method for sustainable water management, which could be applied everywhere like a recipe. Further, water management is linked to earlier development, social and political structures and geographic background and must account for this.

3.1 Definition and characteristics of sustainability

It is popular to reason with sustainability. The World Commission on Environment and Development defined sustainable development as "development that meets the needs of the present generation without compromising the ability of future generations to meet their own needs" (Brundtland report, 1987). According to the Commission of the European Community, criteria of sustainable development are to maintain quality of life, keep up continuing access to natural resources, and avoid lasting environmental damage. Sustainable water management has to ensure, that societies today and future can live without compromising the natural hydrological cycle and ecosystem integrity (UNESCO, 1996). In short: "sustainability avoids future regret for decisions made today" (Geiger et al., 2006). However, all of these definitions are too general in order to be of any help to decision makers in water management.

Towards the end of the last century it was realized, that development focusing on economic growth only will not remove poverty and secure future. It may make some people more rich, but the majority of people poorer. Instead, the three dimensions environment, society and economy must be equally addressed.

There are different opinions, which of the dimension should be dominant and if equal by which measure equity is expressed. In view of sustainability certainly environment is connected most to the development potential of future generations. Therefore the World Conservation Union, the United Nations Environment Program and the World Wildlife Fund for Nature defined sustainable development as a development that improves: "the quality of human life while living within the carrying capacity of supporting ecosystems." (IUCN,UNEP, WWF (1991) extracted from UNESCO N.D.)

This definition corresponds to the very old idea of sustainable forestry, developed by Hans Carl Carlowitz in 1713, an German forestry engineer from Freiberg. He defined a very simple balance which has to be fulfilled to reach a sustainable status for a given forest unit: "wood amount cut in one year ≤ wood amount growing up in one year." At last this is an excellent example of what is meant by the expression "carrying capacity of supporting ecosystem". For the assessment of sustainability it is essential if this principle is applied or not. The following distinction is important for the definition of indicator sets for sustainability evaluation:

- Strong sustainability. Strong sustainability means that preservation of natural capital is considered as not substitutable by any other form of capital and the imbalance is fulfilled for every case.

- Weak sustainability. Weak sustainability implies that the depletion of natural capital can be substituted by man-made capital, as long as the sum of both is not decreasing.

The different components which make water management sustainable did not receive the same priorities at all times. At different stages of societal development, different objectives are considered more important. In the pre-industrial society, emphasis was placed on drinking water supply and irrigation. In the industrial society, generation of hydropower and waste disposal and transport are prioritized. Finally, in the post-industrial society, high emphasis is placed on aesthetics and ecology. While the existence of changing priorities must be recognized, still lower priority uses cannot be neglected over a long run of time, because of the interdependency of various uses.

3.2 General requirements for integrative management

United Nations (UN, 1994) and UNESCO (UNESCO, 1987) through its International Hydrologic Programmes (IHP) in the 70ies of last century tried to gather meteorologists, hydrologists, biologists, social scientists, economists, agrarians, city planners and others to support integrated and holistic approaches. This has enhanced sustainability discussions and development of sustainability indicators especially on global issues. Until today, however, workable regulations to assess sustainability of different management options not exist.

Decisions in water management have to be done on different scales, local i.e. for an individual urban development, regional i.e. for the Shandong coastal area, for river catchments i.e. Huangshui river basin or Yellow river basin, maybe continental i.e. Australia or even global.

Integrated water management recognizes the system complexity. It equally involves local and regional authorities, environmentalists and decision makers, politicians of all parties, governing and in opposition, and especially the people affected. Sustainable water management ensures that no substances are accumulated or energy is lost by recovery and reuse techniques. This is easier said than done for the reasons discussed in chapter 2.

The creation of inter sector links, supporting cross-sector cooperation and integrated multi-disciplinary actions may bridge existing sectarian structures and the new goals for implementation of sustainable water management. Ultimately, sustainable and integrated water management needs novel administrative approaches and holistic education.

In Europe sustainable and integrative principles are included in the European Water Framework Directive (EUWFD), enforced since 2002. However, when applying the European regulations it was found, that conventional procedures for establishing water management plans and even more for selecting measures to implement the plans were not sufficiently taking into account sustainability aspects as requested.

In China on August 29, 2002, during the 27th meeting of ninth people's congress the "water law" was modified, which brought many breakthroughs in water management regulations. The new law is based on the destination of setting up a water saving and waste water prevention society and gave priority to water resources protection and optimization of the compromise between economy, society and ecology. Local governments began to focus on water management and many files were enacted. Currently a tough problem in China is how to overcome the sectarian division of responsibilities and political priorities influencing water management decisions. After all, in Europe like in China the legal background for sustainable water management is provided, but in both cases precise and workable regulations how to proceed with sustainability assessments are lacking.

3.3 Categorization of sustainable water management objectives

The identification of base-level objectives is a very demanding first step within an evaluation system. For a transparent evaluation system which represents all dimensions of sustainability it is necessary to define objectives in respect to one fact or process only and independent to each other. For example,for the environmental dimension the two objectives "minimizing phosphorous discharge" and "minimizing entropic status of lakes" theoretically could be identified. However, phosphorous concentration is the critical parameter for an entropic status of a lake. Therefore these parameters are not independent. If these two parameters would be both used as objectives the entropic status of the lake would be represented in excess within an evaluation. It is a laborious trial-and-error process to define a set of base-level objectives for each individual case, which is comprehensive on the one hand and independent on the other.

The next step is cataloging calculable objectives. To achieve this, indicators must be defined which finally have to be merged somehow for evaluation of sustainability. This suggests a thinking structure staging general, top-level and base-level objectives and related calculable indicators. Chapter 4 will elaborate on indicator systems and evaluation methods in detail.

Same principles as requested for the formulation of top-level objectives apply to indicators for base-level objectives. Neither two or more base-level objectives should be integrated into one indicator nor should more than one indicator be defined for one base-level objective. Otherwise an over- or under-representation of objectives would occur. The question may arise, if finally there should be only one indicator for one dimension permitted.

A problem in all societies is, how to deal with basic needs and with the poor. Social competence must guide new approaches for water supply and sanitation. In mega cities of developing countries infrastructural systems must be designed according to the needs and possibilities of the majority. Universally valid standards may be contra productive in some cases. Public participation prevents social conflicts whereby social development is not self-evident, but must be initiated and guided.

The ability to quantify, in monetary terms, the socio-economic dimension of a water project also does not determine its substance. Analysis should differentiate between price and value in situations where price is a current, transient factor, and not an accurate measure of ultimate value. The conventionally used monetary costs should be supplemented by resource and environmental costs, which consider positive environmental effects as well as environmental damage caused by implementation of environmental measures (UNESCO 1996).

Last not least it should be mentioned that setting objectives and planning by itself will not solve the problems, but helps to identify the most socio-economic compromise to achieve a good status in surface and groundwater. Water management planning is useless, if measures are not implemented and environmental laws are not enforced, i.e. controlling industrial or other emissions and sanction offences. One should think, what Winston Churchill meant, when he said: "At first we form our environment, then it forms us." This may lead to the true objective of water management: "An ounce of precaution is better than a pound of cure."

3.4 Variation of water management objectives in different planning phases

The chapter 3.1 illustrated, that overall objectives change with phases of development of whole societies. In addition one has to realize, that the potential to influence sustainability and thus planning goals are differ in the diverse planning stages(see Table 1).

Table 1 Goals of sustainability evaluation in different planning stages

Planning stage	Goals	Indicators	Comments
Pre-planning	To make the best choice of a system i.e. water supply, storm drainage, waste water treatment, flood protection, surface-groundwater protection	The choice of indicators strongly depends on natural and local conditional as well on stakeholder interests	Here most money can be saved, environment protected best and social concerns considered most. Aside from earlier layouts no data exist
Planning	System choice and environmental protection ways largely are fixed, only technical solutions in more detail can be evaluated	Indicators may address cost efficiency and affordability for society	Still economic savings and environmental protection maybe optimized. Basic systems layout must exist
Design	The choice of material etc. still can be made	Indicator may address costs, durability etc.	Decided layouts are designed, the cost saving potential is quite small
Implementation	Different construction processes may be sequenced best	Different options can be compared for costs, in water sensitive areas for environmental impacts	Savings in any respect are marginal
Operation	Different operations scheme may be compared	Indicators cost and financial burden for the users	Economic saving and environmental issues can be influenced, but by far not as much as in the pre-planning phase

Obviously, sustainability can be influenced most in the pre-planning stage. Here most money can be saved and environmental and social concerns can be considered best. In planning and design stages still economic savings and environmental protection maybe optimized to some extend. During implementation savings in any respect become marginal, while during operation economic saving potential and environmental impacts again may be significant, but by far not as large as in the pre-planning phase.

4 Review of indicator systems

4.1 Review of general indicators on sustainable development

Many different indicators for sustainable development were suggested by various institutions in the past decades. Some were transformed into national criteria of sustainable development and eco-city development in China.

Water Poverty Index

Poverty is closely connected to water management. The Water Poverty Index is an integrated index, including five aspects: resource availability, access to water, capacity of people to manage water, use of water, environment. The main purpose of this index is concerned with water resources and related issues of human poverty.

4.2 Indicator systems for sustainability evaluation on river-basin level

4.2.1 Victorian Catchments' Indicators

Victoria located in the Murray-Darling Basin in Australia experienced consecutive droughts of eight years. The local government began to realize that exercised water management is not sustainable (Ayers, 2003). Relying on dams, reservoirs, seizing large quantities of water to secure human life, industrial development and agricultural irrigation and discharging raw wastewater, increasingly caused problems. Thus, local government has proposed a system to guide people to use water in a sustainable way after analyzing situation and history, five principles for water management were advocated: ①A healthy environment is the basis for social and economic development. ②Government departments should be the starting point for the interests of all the people of Victoria, to make efforts to protect all water resources. ③All people have ownership of water utilities. ④All water users must pay in proportion to their enjoyment the entire cost of water services, including investment in infrastructures. ⑤Victoria's water sector is in charge of the whole community water system.

4.2.2 Indicators applied to Yangtze River in China (China water, 2005)

Yangtze River with a length of 6,300 km, is the longest river in China and the third longest in the world. It's total runoff is $1,000 \times 10^9$ m^3/a. Its basin drains 1,808,500 km^2. The basin supports 36% of China's population and accounts for 40% of food supply, 33% of grain crops, 47% of freshwater products and 40% of GDP (China water, 2005). Due to the huge natural resources, especially the downstream region of Yangtze River Basin observed remarkable development in recent years. Ignorance of sustainability development produced many problems, i.e. deforestation in upstream reaches, provoking big losses of water and soil. The resulting mud silting frequency, floods and droughts in downstream reaches, unsound industrial development causing heavy pollution and subsequent damages on ecology decrease fish resources and increase incidences of schistosomiasis.

Now central government has established a "sustainable development strategy" as key strategy that always should be followed. The healthy Yangtze River indicator system is the first quantified river basin management indicator system in China. There are four levels in this system, general objective level, system level, status level, and factor level.

The general objective is to maintenance the health of Yangtze River and to balance human development and water resources. The system levels are ecological environment protection system, flood control and security system, and water resource development and utilize system. There are five status levels: status of water(soil) resource and water environment, status of the integrated stability of the river, status of aquatic biodiversity, status of flood retention capacity, and status of service capacity.

15 indicators were quantified:

(1)satisfaction degree of ecological water demand in river course.

(2)percentage of water up to par in water function area.

(3)percentage of soil and water loss.

(4)percentage of interdiction on schistosomiasis.

(5)wetland reservation.

(6)maintenance of river status.

(7)fish bio-integrality.

(8)completeness of flood control engineering and no-engineering measures.

(9)development and use of water resource per 10^4 RMB GDP.

(10)safety guarantee of drinking water.

(11)water supply guarantee in towns.

(12)irrigation guarantee.

(13)use of water energy resource.

(14)water depth guarantee for navigation.

(15)living status of rare aquatic animal species, and connectedness of the water system.

4.2.3 Indicators developed for Yellow River in China (Xinhua net, 2004)

Yellow River also spelled Hwang Ho, (Pinyin Huang He, with English Meaning Yellow River), River of China, often called the cradle of Chinese civilization. It is the second longest river of China with 5,464 km and drains the country's third largest basin with an area of 750,000 km^2. The river rises in Tsinghai province on Tibet Plateau and crosses six other provinces and two autonomous regions in its course to the Po Hai (Xinhua net, 2004).

Assessment of the Yellow River condition and research of possible indicators to define the health of the river are ongoing. The general objective is to maintenance life in Yellow River. There may be three pillars in an indicator system, which are: ecological indicators, human health and socio-economic indicators, and physical-chemical indicators.

4.3 Indicators for sustainability evaluation of urban developments

The wide-ranging indicators of chapters 4.2 and 4.3 may be sufficient to assess overall situations and to support government decisions on setting priorities towards sustainable development, but are not very helpful to assist the selection of management alternatives and structural or non-structural measures taken for the many individual developments especially in urban areas. Such indicators must be calculable with a certain precision and have to account for the specific local situation. Some examples are given for costs, water quantity and water quality.

4.3.1 Calculable index to reflect costs

The costs of different scenarios can be evaluated by the "Actual Cash Value" (ACV) of a scenario (MUNLV, 2005). The ACV of all measures belonging to a scenario is added. This method is used to compare measures with different durations of useful live and different ratios of investment and operational costs. The interest rates for loans and the price increase during the evaluation period are considered. The ACV is related to a baseline year, typically the beginning of the operation of the measure. The ACV is the monetary value of a measure corresponding to the purchasing power within this year. This value is much less than the sum of all expenses at the end of the evaluation period, because financing costs will not be calculated if the loan is paid out in the base year.

The following example compares two measures with different ratios of investment and operational costs and same duration of useful live.

If the operational costs are simply multiplied by the duration of useful live and added to the investment costs, the total cost would be 1,300,000 ACV for both measures.

For the calculation of the ACV an interest rate (IR) of 3% and a rate of price increase (RPI) of 1% is used. The ACV will be calculated by the following equation:

$$ACV_{OC} = NV_{OC}(1+IR)\frac{(1+IR)^n - (1+RPI)^n}{(1+IR)^n(IR-RPI)}$$

with, ACV_{OC} is actual Cash Value of operational costs,
and
$$ACV = ACV_{IC} + ACV_{OC}$$
with ACV_{IC} (Actual Cash Value of investment costs)= NV_{IC} (only valid if the hole investment is made in base year).

The calculation yields for measure 1 an ACV value of 827,654 and for measure 2 an ACV value of 680,046. The difference between these two values underlines that a method of cost accounting is needed for an accurate monetary evaluation of scenarios.

4.3.2 Indices for utilization and exploitation of natural water resources

A balance of consumption and regeneration of water resources actually is only comparable on watershed level. This also would apply to urban areas, where landscaped areas and undeveloped parts of the watershed have to be taken into account. For practical purposes also for parts of catchments indicators can be defined for surfacewater and groundwater, which reflect sustainable water use:

$$I_{GW}/I_{SW} = \frac{CO_{GW}/CO_{SW}}{RG_{GW}/RG_{SW}}$$

with, I_{GW} is indicator for groundwater. I_{SW} is indicator for surfacewater. CO_{GW} is consumption of groundwater (during a defined period of time). RG_{GW} is regeneration of groundwater (during a defined period of time). CO_{SW} is consumption of surfacewater (during a defined period of time). RG_{SW} is regeneration of surfacewater (during a defined period of time).

4.3.3 Indices for protection of groundwater

Because of the complex interactions it is impossible to define evaluation marks for different substance loadings to a specific groundwater body. The only practical way to prevent negative effects to the groundwater i.e. by infiltration is to limit infiltrated concentrations. Because generally valid information about the elimination of substances in the unsaturated natural soil are lacking these limits have to be applied to the outflow of infiltration facilities.

For storm water infiltration for instance the following substances may be relevant:
- Cadmium (Cd)
- Copper (Cu)
- Lead (Pb)
- Zinc (Zn)
- Poly Aromatic Hydrocarbons (PAH)
- Mineral Oil Type Hydrocarbons (MOT)

"Environmental Quality Standards for Groundwater of People's Republic of China" (GB/T 14848—93) are containing limiting values for Cd, Cu, Pb and Zn. There five quality classes are defined. Only water, which is classified as class A can be used for every purpose. Because sustainability is defined as the unlimited possibility of development for future generations the limiting values for the class A should be used in such a case.

4.3.4 Index to reflect social issues

A possible social indicator is the financial burden for citizens. A possible indicator could be the total costs for water supply and waste water disposal per citizen and year. Fees, however, have to be payable by citizens and have to be in relation to their income. A better indicator is to set these costs in relation to the income of citizens, because there will be a different burden if these costs is 0.1% or 10% of income. Problems appear in calculating the costs and getting information about the local income.

Another social indicator may be risk. Water supply and waste water disposal at all times should be ensured. Casualties i.e. with contaminated water should be reduced. A possible indicator could be a statistical value of casualties caused by contaminated water.

A high level of health and life quality is also a social indicator. Good environmental conditions increase recreation, life quality and finally health. There are different water quality requirements, i.e. for swimming, fishing, sailing, diving, etc..

Another important social indicator in sustainable water management finally is the participation of different stakeholder in all phases of planning, implementation and operation.

5 Decision support to select sustainable water concepts/BMP's

The following approach tries to accommodate for the deficiencies mentioned in chapter 4.3. The key part of the new system is the selection and calculation of indicators for practical purposes and the rating procedure for an overall sustainability evaluation. Both are based on theoretically sound formula. With the approach presented indicators may be evaluated for different purposes and on different scales.

5.1 Theory and structure for the Hierarchical Decision Support Concept

For the case of different levels of criteria a two stage decision system is suggested. The two stages are selection(combination) criteria and advanced criteria. For stage one it is suggested to establish independent measure catalogues, from which according to pre-designed selection(combination) criteria independent from the user scenarios are established and ranked. Prior to selection(combination), the measures must be checked for there hydrologic, geologic and socio-cultural applicability to the area under consideration. This step is added to the conventional approach to water management as a basic pre-condition. Only a small number of alternatives can be investigated in more detail to lead to a final strategy. This is called decision support system level two.

For decision support level two it is suggested to first establish a GIS based detailed data base which dependent on the objective also can be updated by satellite images of surveys. Starting from this database data for the planning horizon must be extrapolated. Then suitable modeling techniques according to data availability, planning goal, time availability and available finances must be defined. As a next step pre-selected scenarios must be optimized and checked if they can be implemented establishing stage plans and detailed cost estimates.

The final selection of the most cost-efficient solution again similar to decision support level one must be then on the basis of environmental and resource effects checking it again for its social acceptance and financing possibilities.

The advantage of the system is that it can be understood by decision makers. This is not the case with most of the decision support systems which exist so far, where the decision makers have to trust that complex modelling schemes are true. Even more, with the suggested system the input data used for calculations are laid open to the decision maker and himself can judge on the validity on the each and every input value.

5.2 Ranking of indicator values

For ranking of alternatives Lu Zhibo (2006) suggests the "Short board (bucket) theory". The short board theory is popular and universally accepted in the field of economy. According to this theory, the capacity of a wood bucket hoop by wood pieces is not depend on the longest piece or the average piece; it is decided by the shortest piece. If one wants to increase the capacity of the bucket, the length of the shortest piece must be improved.

This theory is also fits sustainable development issues. Local sustainable development of a city depends on natural resources, technologies, labour force, environment, financing, etc.. No city can have all the advantages of sustainable development. Sustainability is like the wood bucket and is decided by the shortest constraint. Evaluation methods should concentrate on the shortest factor. For more details see Lu Zhibo (2006).

6 Conclusion

Water resources management today is much more than applying mathematical process models, producing GIS-based maps or designing, implementing and operating technical schemes. It must consider environmental needs and prevents social conflicts by integration of stakeholders into the planning process. This achieves sustainable water management.

The obstacles to achieve integrated water management are seldom isolated ones. More commonly they are linked with each other, although connections may not be recognized. It also is recognised that the traditional ways of water resources managements must be markedly modified in order to cope with these deficiencies and with the continuing increase of water problems. It is clear that today's water problems only can be tackled by efficient and well-experienced water resources management bodies having sufficient political support.

The presented two-stage Decision Support System is able to integrate technical, social and political issues for water management decisions. This kind of approach actually is envisaged by Chinese law and some regulations. The suggested decision support system could bring a change from the traditional "Water management by nine dragons" to "Water management by one dragon". However, it only can succeed, if administrative changes are made. A cross-sectarian association could take charge of the integrated functions of flood control, waterlog removal, drought fighting, water supply, drainage, water saving, waste water treatment, recycling, rain water harvesting, ground water recharge, watercourse remediation, and others.

References

[1] Adler and Ziglio. Gazing into the oracle, Jessica Kingsley Publishers:Bristol, P.1996.

[2] Ayers G. CSIRO Climate and the Murray-Darling Basin, MDBC Climate Change Workshop, 2003.

[3] Chinawater. Yangtze River water authority, "Protection and development Yangtze River Pronouncement",2005.

[4] http://www.chinawater.net.cn/zt/changjiang/huiyibd.asp?CWSNewsID=19156, 2005.

[5] Cornish. The study of the future, World Future Society: Washington D.C., 1977.

[6] EU-WFD. EU Water Framework Directive 2000/60/EC of the European Parliament and of the Council establishing a framework for the Community action in the field of water policy", 2000.

[7] Geiger et al.. Cost-effective Measures for Flood Control and Groundwater Recharge in coastal catchments. Research project founded by BMBF, Förderkennzeichen: Bonn, Germany 2005, CHN 04/015.

[8] Geiger, Lu Zhibo and Lu Yongsen. Evaluation and indicator system for sustainable water management. UNESCO course, 2006.

[9] Helmer. Problems in futures research: Delphi and causal cross-impact analysis, Futures, PP. 17-31, February 1977.

[10] Lu Zhibo. Study and theories and practices of regional sustainable water management indicator system. Tongji University, dissertation, 2006.

[11] MUNLV. Method for the Evaluation of Pollution Loads from Urban Areas at River Basin Scale for the identification of significant human pressures and impacts on water bodies as required by EU-WFD, 2005.

[12] Nafo et Geiger. Assessment of the significance of nitrogen loads from point and diffuse sources in a small river basin, IWA Conference Marrakesch 2004 (accepted for oral presentation, peer review), 2004.

[13] Nafo. Bilanzierung zur Beurteilung von Niederschlagswassereinleitungen auf regionaler Ebene. Schriftenreihe Forum Siedlungswasserwirtschaft und Abfallwirtschaft der Universität Essen, Heft 23, Diss, 2004.

[14] UNESCO. Manual on Drainage in Urbanized Areas, Volume I: Planning and Design of Drainage Systems, Volume II: Data Collection and Analysis for Drainage Design, Vol. I + II, W. F. Geiger, J. Marsalek, W. J. Rawls, F. C. Zuidema, 1987.

[15] UNESCO. Integrated water Management in urban and surrounding areas. Contribution to UNESCO-IHP project M3-3a, 1996.

[16] United Nations. Guidelines on Integrated Environmental Management in Countries in Transition. New York, 1994.

[17] Russ. Bausteine einer nachhaltigen Siedlungsentwicklung und Ihre Berücksichtigung im Wiener Strategieplan, Wien, 2000.

[18] Sulivan et al.. Sullivan C.A., Meigh J.R., O'Regan D. Evaluating Your Water a Management Primer for the Water Poverty Index. Wallingford, Centre for Ecology and Hydrology, 2002.

[19] Wissema. Trends in technology forecasting. R & D Management, 12(1), pp. 27-36, 1982.

[20] World Commission on environment and development. Our Common Future (Brundtland Report), Oxford University Press, 1987.

[21] Xinahua net. Yellow River water authority, "How to sustain the health Yellow River", 2004.

[22] http://www.ha.xinhuanet.com/xhzt/2004-06/09/content_2281849.htm 2004.

Methodology and Practical Examples of Model-based Groundwater Monitoring, Management, Protection and Remediation

Stefan Kaden

(WASY GmbH, Institute for Water Resources Planning and Systems Research, Berlin, Germany)

1 Introduction

Groundwater is an important resource for water supply and at the same time an important component of our aquatic environment. Many countries are entirely dependent on the groundwater resources. The threat to the sustainable use and management of groundwater is immense. It is frequently at risk from depletion and pollution.

Groundwater systems as hidden and "slow" components of the hydrological cycle are usually characterized by large time gaps between cause and consequences of systems changes. This is especially true for negative impacts and for remediation measures. Consequently there is a need for long-term, preventive monitoring, groundwater protection and remediation.

In many areas, there is a lack of information on the position and potential effect of dominant pollution sources within the catchment area of the waterworks. Many of the effects are first noticed in the future when large quantities of water are already contaminated and remediation measures become extremely non cost-effective and cumbersome. Therefore it is highly important to gain information on the potential risk from different contaminated sites as early as possible and in a structured and comprehensive way, as well as to forecast potential risk and for planning of remediation measures. This is one task for numerical groundwater modeling.

Other groundwater modeling tasks relates to water resources planning and management, e. g. the design of groundwater recharge systems, the estimation of well-head protection zones or the integrated groundwater - surface water management in agricultural areas.

In the last decade, groundwater modeling has changed from an expert tool, applied mainly by specialists to special problems, to a widely distributed instrument in groundwater hydrology, water management and environmental protection. This development was forced by advances in computer systems and modeling techniques, as well as by new challenges in practical applications. The most obvious example of new achievements is the 3D modeling of coupled groundwater flow, mass and, in part, heat transport processes. At the same time new tools in pre-processing and post-processing have been developed and implemented in order to handle large, complex tasks. Consequently, 3D-groundwater modeling has become almost common practice.

Nevertheless, it has become apparent, that 3D modeling of coupled groundwater flow and mass transport under real-world conditions is more an art than a craft.

The paper first presents basics of groundwater modeling technology. After that the interrelationship between groundwater modeling and monitoring will be discussed.

Finally three modeling examples will be presented.

The first study concerns the planning of a sustainable, integrated land and water management for a low land along the Odra River in Eastern Germany. In this case, the integrated monitoring and modeling of surface water and groundwater was most essential.

In the second case study possibilities and limits of artificial groundwater recharge using storm water within the Olympic Park Beijing 2008 have been analyzed.

The third study concerns the planned new International Airport Berlin-Schönefeld. One of the problems in the planning phase was the large number of known and anticipated hazardous wastes, mainly from Second World War. A 3D groundwater flow and mass transport model was used to evaluate and improve an existing groundwater monitoring program and to plan pollution remediation measures.

2 Modeling technology

(1) Data collection, acquisition and processing.
(2) Information-adequate (conceptual) model building.
(3) Numerical model building.
(4) Model application, interpretation of model results.
(5) Final data storing and result preparation.

3 Monitoring and modeling

Groundwater monitoring and groundwater modeling are closely interrelated. On the one hand, groundwater modeling related to practice is not thinkable without monitoring; on the other hand, groundwater modeling can contribute to a large extend to improve the significance of any groundwater monitoring, even reducing costs for that. This interdependency is illustrated.

Groundwater monitoring serves for the description of the properties and state of the groundwater system until "today". Otherwise, the main application area of modeling starts with the present state with forecasts and scenario analyses. Modeling of historical states serves for model calibration and verification as well as the analysis and interpretation of groundwater processes. Scenario analyses otherwise serve the improvement of groundwater monitoring concepts. This interplay between modeling and monitoring may be described as the concept of "permanently working modeling" or "site modeling". Necessity and obviousness of the interplay becomes clear, taking into the account that any groundwater model is just an abstraction of real world. By the help of permanent, comparative interpretation of modeling results and monitoring, this abstraction can be improved and consequences of this abstraction can be interpreted.

4 Modeling for sustainable land and water management

4.1 History, location and Objectives

The Oderbruch, a large polder area along the Odra River in Brandenburg, covers an area of 825 km^2. It is located in the eastern part of Brandenburg and was a frequently flooded flood plain area until the 18th century. Agriculture was only possible in the higher part of the Oderbruch. From the beginning of the 18th century, the Oderbruch has been drained by a multiplicity of ditches and some pumping stations. A main condition for the use of agricultural land was the construction of a dam between the higher located Odra water table and the flood plain area.

Consequently, the Oderbruch is traditionally characterized by an integrated land and water management over centuries. New socio-economic development in the region, new attempts in water resources management and nature preservation and last, but not least, the European Water Framework Directive call for new sustainable approaches in land and water management. These approaches have to take into the account the partial contradictory interests of different stakeholders in the region.

In the framework of a development program of the state of Brandenburg to improve the landscape water balance as well as the rural environment, a number of so called "agro structural development plans" have been worked out. One of these plans, the plan for the Glietzener Polder, will be introduced in the following.

The Glietzener Polder is a 112 km^2 large sub area of the Oderbruch and is bounded by the great receiving rivers Odra in the east, Alte Odra in the south and Stille Odra in the north. The polder is drained by eight pumping stations.

The Odra River forms the decisive boundary condition with a water table above the land surface of the polder. Within the polder a dense system of ditches has to be considered. To model the important interaction between surface and groundwater a model system has been developed with FEFLOW[1] (Diersch, 2005) for 3D groundwater flow and MIKE11[2] for hydrodynamic flow. For details on the coupling see Monninkhoff (2003).

4.2 Groundwater model

In the course of different investigations concerning water management in the Oderbruch in the years 1996 to 2000 by WASY GmbH, a complex groundwater model was developed (LUA, 2002). The model was developed with FEFLOW, a 3D finite element modeling system. This calibrated groundwater model formed the basis for a series of investigations regarding water resources planning and management in the Oderbruch and also for the current investigation for the Glietzener Polder.

From the overall groundwater model of the Oderbruch a detailed model of the Glietzner Polder was derived. This model was built up as a transient model (for a typical hydrological year), considered the inner annual variation of boundaries as the water tables in the river system as well as the natural groundwater recharge.

The importance of considering the processes as transient, results among other things from the range of oscillation of the Oder's water table.

The hydrogeological conditions are characterized by a shallow sandy aquifer covered widely by very low permeable alluvial sediments. The sediments are in some cases faulted by clay congregations. For all dewatering ditches, a boundary condition of the third kind was use. Depending on the distribution of the alluvial sediments, the boundary condition was adapted corresponding to a weakly permeable layer. The Odra as the main receiving stream was implemented by a boundary condition of the first kind (given water table). In order to analyze the water tables around the partly small ditches and pumping stations, the finite-element-mesh was refined around these special zones. The mesh consists of 110,000 elements divided into four layers.

4.3 Hydrodynamic model

The model domain including all cross sections, weirs and culverts is shown in Figure 1.

4.4 Calibration

Figure 2 shows the results of the groundwater measurements fragmented into months.

As result, a high quality model was made available for the detailed investigation of the current state of water management in the Glietzener Polder and the following planning analysis.

---

[1] FEFLOW is a registered trademark of WASY GmbH.
[2] MIKE11 of DHI Water & Environment.

4.5 Planning steps
4.5.1 Definition of targets—Agriculture

Agriculture has defined monthly groundwater levels for farmland and meadow areas depending on land use. Other targets have been defined above others for nature protection.

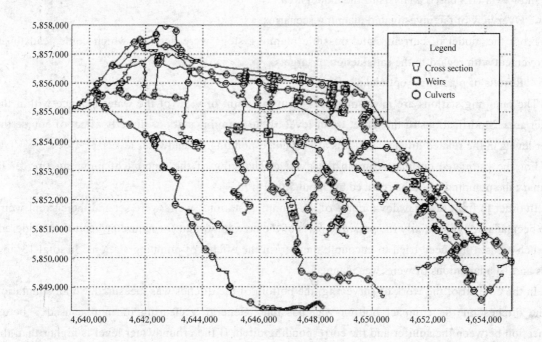

Figure 1   Model domain Glietzener Polder

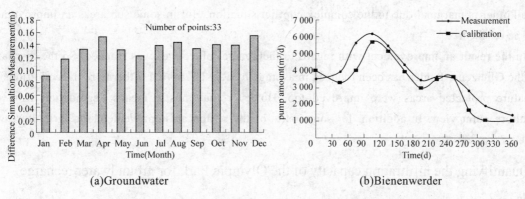

(a)Groundwater (b)Bienenwerder

Figure 2   Calibration statistic for groundwater and pump amount

4.5.2 Landscape water balance

The main aim from the point of view of ecological oriented water management is the reduction of the water discharge out of the study area, i.e. to keep as much water as possible within the region. On the one hand, the groundwater level will be raised and resulting from that operational cost of the pumping stations will be reduced.

The different targets show the large conflict potential. A high conflict potential exists especially between agriculture and nature conservation and between agriculture and landscape water balance, respectively.

4.5.3 Analysis of the current state

The current state has been analyzed concerning the different targets by the help of the model system (modeling the selected typical hydrological year). The water characteristics of the model area differed between "too dry", "too wet" and "acceptable". The features refer to the groundwater level. Was show by a GIS based analysis of the model area.

4.5.4 Overview of the investigated planning scenarios

From the model of current state, possible planning steps were derived, which could lead to an improvement with regard to the satisfaction of targets.

4.5.5 Results of planning—optimal scenario

The pumping stations are most important from the point of view of the water balance within the Polder area. Modifications of the target water level of the pumping stations directly affect of course the water levels in the related ditches. And this affects corresponding groundwater levels.

For the estimated optimal scenario the average changing of the amount of pumping (m^3/d). In summary the pumping could be reduced by about 5,000 m^3/d.

In order to limit the drawdown range of the pumping stations, weirs were set and the current weirs have been adapted. Especially the sensitive parts of the area important for nature protection considered. For such areas it has been tried to uncouple them form the effects of pumping stations. In total 13 new weirs and 13 adaptations of weirs was planned.

In the context of the work of surveying, the bottom of the ditches was measured. So an adaptation of the ditches on the compact bottom was possible. General a clearing of mud caused a better connection between the aquifer and the corresponding ditch. If the groundwater level is higher than the ditch water level clearing of mud causes a drawdown of the groundwater level and a increasing of the discharge. In total ditches with a length of 13.3 km have been planned for clearing of mud.

The percentage change in the satisfaction of targets of nature conservation and agriculture in the whole Polder is marginal due to the complex conflict situation. But in some sub areas an improvement could be reached.

In the result an improvement with regard to the general objectives was reached. So the discharge from the Glietzener Polder has been reduced in average by 1.7×10^6 m^3/a. Furthermore the conditions for the nature protected areas were improved too. However that mostly means degradation from the agriculture point view. In addition, for some parts of the region an improvement for agriculture was reached.

5 Quantifying the infiltration capacity of the Olympic Park for groundwater recharge

5.1 Introduction

The Chinese-German Project "Sustainable water concept and its application for the Olympic Games 2008", supported by BMBF and MOST, intends to apply the latest scientific results, methods and technologies for a sustainable water management of the Olympic Park. Within the sub-project of WASY possibilities of groundwater recharge with storm water have been analyzed. Some results are presented in the following.

One planning alternative for a sustainable water concept of the Olympic Park was to recharge groundwater via injection wells. If the underground allows sufficient infiltration an appropriate concept for artificial recharge should be developed.

The first step was to set up a hydrogeological model. For that purpose the software HydroGeo

Analyst (Waterloo Hydrogeolgic Inc.) was used. Based on the hydrogeological model a 3D groundwater model with the software FEFLOW was developed for the estimation of the infiltration capacity and calculation of different scenarios.

5.2 Creating a Geometric Model

According to available information a regional model area has been selected with about 39 km^2.

The first step of the model construction is the spatial subdivision. Therefore a supermesh was generated, which contains important topographically and other structures and consists of 54 polygons. Based on this the Finite Element Mesh was developed with 73,566 triangular elements and 43,463 nodes relating to 6 model layers and 7 slices.

5.3 Designing a hydrogeological model

The planned Olympic Park is situated at the down part of the alluvial fan of Yongding River, going down from south to north topographically. In the central area of the Olympic Park normally there are two to four aquifers in a range of 40 m, typically three aquifers.

The first aquifer extends into a depth of 10 m and consists of fine sand and silt. This belongs to the type of the "tableland aquifer" (unconfined groundwater). The second sandy aquifer is between 10 to 25 m under ground surface, belonging to the interlayer aquifer (interlayer groundwater). The third aquifer is below 35 m, consisting of sand and gravel, belonging to the confined to unconfined kind of main aquifer.

For the central area of the Olympic Park, a lot of borehole data and geological cross section have been provided. Outside of this area only a few information existed. Therefore it was necessary to interpolate and extrapolate respectively for the preparation of the hydrogeological model. The available boreholes in the whole model area and give some information about the depth and the number of boreholes and created cross sections with HydroGeo Analyst. Only a few boreholes (approximately 10 boreholes) were so deep that they reached the confined and unconfined 3rd aquifer. Most boreholes intersect only the 1st and 2nd aquifer.

For the central area of the Olympic Park, the knowledge of the hydrogeological situation about the first 30 to 40 m depth is very good. With HydroGeo Analyst the following layers based on the available drilling logs could be considered:

(1) Aquitard 1 contains artificial filling materials and top soil most silt and silty clay.

(2) Aquifer 1 corresponds to a powdery fine sand layer and a sandy silt layer.

(3) Aquitard 2 corresponds to silty clay layer.

(4) Aquifer 2 consists of powdery fine sand too and rarely of gravel.

(5) Aquitard 3 is a silt layer.

In the following chapter all important information about the hydrogeological situation and parameterization of the groundwater model are summarized.

5.4 Parameterization of the Groundwater Model

Information about the permeability was given in the reports of Beijing Geotechnical Institute. They were from Pumping tests, Infiltration tests and Lab tests. Partly the permeability values are very different depending on the kind of test. Therefore in the reports values were suggested. These suggested values were used for the generation of the groundwater model. For layers without any information about the permeability values were estimated based on the description in the drilling logs.

In Table 1 all important facts about the geometric model and conductivities are given. It shows a cross section from North to South through the whole model area. For the bottom of the third aquifer no results from the hydrogeological modeling existed because there were no boreholes which intersect the bottom. Therefore the layer thickness was supposed with an estimated value of 15 m.

Table 1 Summary of the model set up and parameterization

No. of Layer	Litho logy	Function	Thickness (m)		Conductivity (10^{-4} m/s)	
			Central Area	Forest Park	Central Area	Forest Park
1	On top soil, below silt	Aquitard 1	1 ~ 8	8 ~ 30	0.005	0.010
2	Fine sand to silt, seldom gravel	Aquifer 1	1 ~ 6	6 ~ 10	0.200	1 ~ 5
3	Silt, seldom clay	Aquitard 2	2 ~ 10	5 ~ 25	0.001	0.001
4	Fine sand to silt	Aquifer 2	4 ~ 5	5 ~ 45	0.1 ~ 0.5	1 ~ 5
5	Silt, seldom clay	Aquitard 3	5 ~ 25	15 ~ 30	0.001	0.001
6	Coarse sand, gravel	Aquifer 3	15	15	5	5

For the estimation of short term infiltration rates only the upper aquifer and accordingly depth to groundwater is relevant. Therefore in foreground of the groundwater modeling stood the realistically reproduction of the groundwater levels of the first aquifer. But the upper two aquifers contain only perched water. For the reproduction of perched groundwater an unsaturated modeling is necessary, which requires a very finely resolved model. Therefore a simplified method of resolution was chosen.

It was only setting a head boundary along the western and eastern borders for the first aquifer based on the ground surface because the water courses along the model boarders no water levels are given. The rest of the lower layers were assumed as saturated.

The water courses inside the Olympic Park did not consider because most have an artificial and usually closed riverbed without a connection to the aquifer. For the northern and southern boarders a non-flow boundary was defined.

5.5 Results

For a first estimation of possible short-term infiltration rates in the central area of the Olympic Park along both sites of the water course were positioned each with 10 infiltration wells. These were realized that water can infiltrate so long in a well until the depth to groundwater is not less than 1.5 m. This depth to groundwater was defined as security for the surrounding properties.

The location of the infiltration wells as well as the difference of groundwater tables between the variants with and without infiltration wells after 5 days in the first aquifer is shown. During the definition of the well locations it was considered that the locations were appropriate as outside as possible of high-density areas. After 5 days the local increasing of groundwater table at the well location is about 4 to 5 m. The increasing range of groundwater table is between 30 to 100 m.

In consideration of made assumption a water quantity of approximately 7,500 m³ would infiltrate within 24 hours.

In Table 2 the infiltration capacity and the resulting water volume related to an area of 1 km² is represented.

Concluding one can estimate that the hydrogeological conditions are not very adapted for artificial recharge of rapid precipitation.

Table 2 Calculated infiltration capacity of the first model calculation

Time	Potential infiltration capacity (m³)	Equivalent water volume on 1 km² (mm)
1 hour	356	0.4
2 hours	681	0.7
6 hours	1,962	2.0
12 hours	3,826	3.8
24 hours	7,439	7.4

These are not very convenient properties for the planned intention. Clearly larger storage capacities would be necessary to infiltrate remarkable amounts of precipitation.

6 Conclusions

In this paper the interdependency between groundwater monitoring and groundwater modeling has been discussed. Just as Yin and Yan both belong together. There is no real groundwater model which is not based upon monitoring results. And for complex groundwater management, planning and protection studies, monitoring have to be optimized based on modeling.

The best results can be achieved, if groundwater monitoring and groundwater modeling are realised not only once in the planning phase of a protect, but permanently during planning and operation. This results in the improvement of monitoring, modeling and finally optimized measures for groundwater management, protection and remediation.

References

[1] Beijing Geotechnical Institute. Hydro-geological Exploration for the Central Area of the Olympic Park, Beijing Geotechnical Institute 30.09. 2005.

[2] Beijing Geotechnical Institute. Hydro-geological Exploration for the C-zone (sinking garden) of the Central Area. Beijing Geotechnical Institute 29.09. 2005.

[3] Diersch H.J.. Grundwasser simulations system FEFLOW: User-/reference Manual, WASY Gesellschaft für wastserwirtschaftliche Planing und Systemforschung mbH, Berlin, 2005.

[4] Kaden S., Fröhlich K., Büttner H., et al. (2001): Grundlagen für die wasserwirtschaftliche Rahmen-und Bewirtschaftungsplanung im Oderbruch, Studien und Tagungsberichte des Landesumweltamtes Brandenburg, Band 31, LUA Brandenburg, 2001.

[5] Monninkhoff B.. Coupling of the groundwater model FEFLOW® with the hydrodynamic model MIKE®, Deutsch-Chinesische Fachtagung, Moderne Methoden und Instrumentarien für die Wasserbewirtschaftung und den Hochwasserschutz", Dresden, IWU-Tagungsberichte, 2003, S. 161-173.

Impact of Stormwater Recharge on Blue Lake, Mount Gambier's Drinking Water Supply

Joanne Vanderzalm[1]   Tara Schiller[2]   Peter Dillon[1]   Stewart Burn[2]

(1 CSIRO Land & Water, Glen Osmond 5064, Australia
2 CSIRO Manufacturing & Infrastructure Technology, Highett 3190, Australia)

1 Introduction

The city of Mount Gambier is located in the southeast corner of South Australia (145° 45′E, 37° 50′S), 445 km southeast of Adelaide and 460 km west of Melbourne. Since its settlement in the 1840's, Mount Gambier has grown to a city with approximately 23,000 residents. Dairy, softwoods, transport and tourism industries, and the factories and services needed to support these, are important to the city's economy.

The city's potable water supply is the Blue Lake, a volcanic crater lake fed from the karstic Gambier Limestone aquifer. Not only is the quality of the lake important to the region for its potable water supply, but its vibrant blue colour in summer is a valuable tourism asset (BLMC, 2001).

Stormwater disposal in Mount Gambier is achieved via sinkholes and drainage bores (more than 500) directly into the Gambier Limestone, thus potentially recharging Blue Lake. As a result, the long-term impact of such stormwater discharge on the city's drinking water supply and the environmental health of the lake is an important management issue for the region. A current research program is assessing the impact and sustainability of this stormwater disposal, given the complex hydrogeological setting of a karstic aquifer system. This paper outlines the methodology proposed to address this issue and the preliminary work undertaken.

2 Background

The geological units that encompass the Mount Gambier region include Cainozoic sediments of the Dilwyn Formation, the Gambier Limestone, Bridgewater Formation and Holocene volcanic deposits (Lawson et al., 1993). This work focuses on the Gambier Limestone, the important unconfined aquifer system in the region and the principal source of water for Mount Gambier. The geology within the heterogeneous Gambier Limestone consists of grey to cream bryozoan calcarenite with thin intervals of marls. Recent work of Hill and Lawson (in press) has identified alternating beds of bryozoal limestone and marl overlying a lower dolomitic Camelback Member. The stormwater drainage bores target this dolomite, intersected at approximately 10 m and 80 m below ground surface, and are as close as 200 m from Blue Lake. A confining layer of glauconitic and fossiliferous marls and clays separates the Gambier Limestone from the confined sandy Dilwyn Formation.

The unconfined aquifer has characteristics of a dual porosity media with both primarily developed porous medium and secondary developed karstic flow. Due to the karstic nature of the aquifer, pump test derived transmissivities vary over two orders of magnitude from 200 m^2/d to greater than 10,000 m^2/d, with porosity estimated between 30% and 50% (Love et al, 1993). Groundwater flow to Blue Lake includes flow through the porous matrix of the aquifer and preferential flow through karstfeatures.

The regional groundwater flow direction in both the unconfined and confined aquifer is toward the southwest, with groundwater discharges via coastal and offshore springs. However in the vicinity of the lake the hydraulic gradient is slight and local flow directions depend on karsticfeatures. The major structural features are related to rifting and post-depositional tectonics, forming a northwest-southeast structural trend through the region.

3 Methodology

The Hazard Analysis and Critical Control Points approach to managing health and environmental risks adopts concentration targets in Blue Lake. The concentration targets for drinking water are based on both potable water quality guidelines (NHMRC and ARCANZ, 1996) and aquatic ecosystem water quality criteria (SA EPA, 2003). As ecosystem maintenance targets for Blue Lake are not well defined, aquatic ecosystem water quality criteria have been adopted at this stage (SA EPA, 2003).

An outcome of this approach is to identify whether water quality is likely to approach two guidelines values: one for drinking water quality and the second for aquatic ecosystem protection, and hence to determine the costs of interventions to achieve different guidelines. Control points include control of contaminant sources, measures to prevent them entering stormwater system and well head traps to improve the quality of stormwater recharging the aquifer. In the case of public health protection, enhanced treatment of water supplies drawn from the Blue Lake is a further control option if required. A related project is assessing structural and non-structural solutions to reducing contaminant loadings entering drainage wells. The value in undertaking both studies is to enable wise investment of funds and effort to achieve demonstrated effective risk management and to increase confidence with which the future quality of Blue Lake can be predicted.

The specific methodology proposed for the risk assessment framework has been initiated with the following steps (to be discussed within this paper):

● Assess and quantify potential contaminants within stormwater.

● Document the possible attenuation mechanisms for these contaminants within the Gambier Limestone aquifer and the Blue Lake.

● Assess residence time within the aquifer.

● Compare the likely concentration after attenuation in the given residence time, with the appropriate concentration target.

Figure 1 shows that for any given stormwater constituent or parameter.

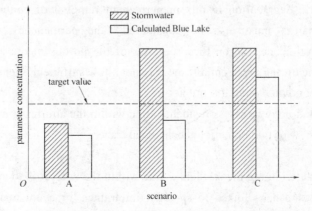

Figure 1　Possible scenarios for long-term water quality impacts on Blue Lake of stormwater drainage to the Gambier Limestone

Scenario A—stormwater concentration is at an acceptable level when compared to concentration target and further attenuation in the aquifer is not required.

Scenario B—stormwater concentration breaches the concentration target for Blue Lake, but with the natural sustainable rate of dilution and attenuation in the aquifer, Blue Lake water quality is expected to comply with the guideline value.

Scenario C—stormwater concentration breaches the target value for Blue Lake and the sustainable rate of dilution and attenuation in the aquifer is unlikely to enable Blue Lake water quality to remain below the guideline value.

Effort in this study will focus on contaminants that correspond with scenarios B and C. To distinguish between cases B and C, information will be acquired on sustainable dilution and attenuation rates of those contaminants within the aquifer.

To date, this work is based predominantly on the assessment of existing water quality data for the Blue Lake, the city of Mount Gambier stormwater and groundwater in the Gambier Limestone unconfined aquifer. A sampling and analytical program has been developed to supplement and extend the existing data. A stormwater sampling program, including frequent sampling during stormwater flow events and passive sampling to estimate total organic contaminant loadings, is currently underway and the initial results of this program will be presented. In addition potential methodology for examining aquifer residence time is being evaluated.

4 Results and Discussion

4.1 Potential contaminants within stormwater

Stormwater exhibits variable quality. Factors influencing this quality include the catchment characteristics, such as the impervious surface coverage and potential contaminant sources, and also temporal variability in contaminant loadings. When considering the potential contaminants within the stormwater of Mount Gambier, it is necessary to consider the dominant industries in the region, typical stormwater contaminants and existing water quality data. Potential contaminants can be compared with the Australian drinking water quality guidelines (NHMRC and ARMCANZ, 1996) and the SA EPA (2003) water quality criteria to identify the potential for scenarios A, B or C.

4.2 Attenuation mechanisms for potential contaminants

Potential attenuation mechanisms within the saturated zone include biodegradation, precipitation and adsorption. These are also possible within the Blue Lake, in combination with photodegradation and volatilization. While degradation results in permanent removal of contaminants, the effects of by-products of degradation must also be considered. The permanency of physical processes, precipitation and adsorption, depend on the properties of the porous media and the contaminant. In addition, the contact time during preferential flow may not allow sufficient attenuation.

4.3 Assessing residence time within the aquifer

Currently the methodology to assess residence time within the aquifer is under development. Thus this section covers a review of two potential geochemical tracers.

4.3.1 Event markers

One approach for determining the residence time within the aquifer is to consider tracers that may be found in the lake that can be linked to specific timeframes, or event markers. Thus, we are not necessarily looking to trace specific sources to the lake, but instead looking for evidence within the lake of anthropogenic contamination which may be linked to certain intervals. Suitable tracers must be relatively conservative to persist to the Blue Lake and also indicative of a specific interval of time. If we consider that the residence time within the karstic aquifer may be of the order of months, it allows for some innovative tracer work.

Organic iodine compounds are being considered as event markers for this study. Organic iodine compounds include triiodinated benzene derivatives, employed as X-ray contrast agents. In this case, the pathway for organic iodine to enter the saturated zone is via leaky sewers as opposed to stormwater discharge. These X-ray contrast media have been in use in Australia since around the 1950's, with details of the specific compounds used in Mount Gambier being sought. There is an organic halogen (measured as adsorbable organic halogen-AOX) signature within the Blue Lake. However, it has yet to be speciated into the contribution from AOCl, AOBr and AOI. While not all environmental AOI can be definitely linked to specific triiodinated X-ray contrast media, it is an indicator that these species are likely to be present. A positive identification of X-ray contrast media within the Blue Lake would indicate recharge post 1950's, or an aquifer residence time of approximately 50 years.

4.3.2 Applied tracers

An applied tracer test, utilising sulphur hexafluoride (SF6) is proposed for use at Mount Gambier. SF6 tracer experiments have been successfully applied to define groundwater travel times near recharge enhancement sites in California (Clark, 2002). SF6, a harmless, non-toxic gas, can be used to 'tag' large volumes of water and is detected at extremely low concentrations, which allows groundwater travel times to be determined accurately.

A field experiment at Mount Gambier will introduce a volume of water tagged with a known concentration of SF6 into selected stormwater drainage wells. This tagged water then has the potential to pass through the saturated zone to reach Blue Lake. While the saturated zone residence time is likely to vary considerably between matrix and fissure flow, this experiment is designed to estimate the minimum likely residence time in the saturated zone, resulting from fissure flow. Thus, the subsequent SF6 monitoring program within the Blue Lake will begin with weekly to fortnightly sampling intervals for the first year.

5 Conclusion

At this stage, a detailed program has been designed to evaluate the impact of stormwater discharge on Mount Gambier's drinking water supply to be completed prior to the end of 2005. The initial field program underway is designed to quantify the stormwater contaminants and to estimate the minimum saturated zone residence time available for attenuation.

After the initial screening of potential contaminants, phosphorus, aluminium, zinc and dissolved organic carbon presently follow scenario B, where the stormwater concentration breaches the water quality criteria (for aquatic ecosystem protection) for Blue Lake but the natural attenuation in the aquifer leads to compliance for Blue Lake water quality. Nonetheless, these species remain of interest to determine the long-term sustainability of these attenuation processes. The current data for chromium suggests scenario C, where the stormwater concentration breaches the target value for Blue Lake and attenuation in the aquifer is unlikely to enable Blue Lake water quality to remain below the aquatic ecosystem protection target value (note that this is not a drinking water quality issue). In this case, the generic target value may not be achievable or applicable to Blue Lake. Further analysis of trace organic species is required to evaluate their potential impact on the Blue Lake.

Overall, this approach allows assessment of the long-term impact of stormwater contaminants posing the greatest risk to the drinking water supply of Mount Gambier and the ecological status of Blue Lake.

6 Acknowlegements

This work is part of the European Commission project Assessing and Improving the Sustainability of Urban Water Resources and Systems (AISUWRS). We acknowledge financial and in-kind support from the Department of Education, Science and Training (DEST), South Australian Environmental Protection Authority, South Australian Department of Water, Land and Biodiversity Conservation, Mount Gambier City Council, South Australian Water Corporation and South East Catchment Water Management Board.

References

[1] Australian Runoff Quality (ARQ). Proceedings of the Australian Runoff Quality Symposium, Albury, Australia, 16-17 June 2003, Institute of Engineers Australia.

[2] Blue Lake Management Committee (BLMC). The Blue Lake Management Plan, June 2001.

[3] Clark J. F.. Defining transport near ASR operations using sulfur hexafluoride gas tracer experiments in P. J. Dillon (Ed.) Management of Aquifer Recharge for Sustainability-Proceedings of the 4th International Symposium on Artificial Recharge of Groundwater, ISAR-4, Adelaide, South Australia, 22-26 September 2002, Swets and Zeitlinger, Lisse, pp. 257-260.

[4] Dillon P. J.. An evaluation of the sources of nitrate in groundwater near Mount Gambier, South Australia, CSIRO Water Resources Series No. 1, CSIRO, Australia, 1998.

[5] Fam S., Stenstrom M. K. and Silverman G. (1987), Hydrocarbons in urban runoff, Journal of Environmental Engineering, 113(5), 1032-1046.

[6] Gonzalez A., Moilleron R., Chebbo G. and Thevenot D. R.(2000). Determination of polycyclic aromatic hydrocarbons in urban runoff samples from the "Le Marias" experiment, Polycyclic Aromatic Hydrocarbons, 20, 1-19.

[7] Hill A. J., Lawson J. S. (in press) Geological setting for the groundwater resources of the lower South East, SA Department of Water, Land and Biodiversity Conservation.

[8] Lamontagne S. and Herczeg A.. Predicted trends for NO_3-concentrations in Blue Lake, South Australia, CSIRO Land and Water, May 2002.

[9] Lawson J, Love A. J, Aslin J.and Stadter F.(1993), Blue Lake hydrogeological investigation progress report no. 1−Assessment of available hydrogeological data, Department of Mines and Energy Report book 93/14.

[10] Love A. J., Herczeg A. L., Armstrong D., Stadter F.and Mazor E.(1993), Groundwater flow regimewithin the Gambier Embayment of the Otway Basin Australia: evidence from hydraulics and hydrochemistry[J]. Journal of Hydrology, 143, 297-338.

[11] Makepeace D. K., Smith D. W. and Stanley S. J.(1995), Urban stormwater quality: summary of contaminant data, Critical Reviews in Environmental Science and Technology, 25(2), 93-139.

[12] Mustafa S., Lawson J. S.. Review of tertiary Gambier Limestone aquifer properties, lower South-East, South Australia, Department of Water, Land and Biodiversity Conservation Report 2002/24.

[13] NHMRC and ARMCANZ. National water quality management strategy: Australian drinking water guidelines, NHMRC & ARMCANZ, Australia.

[14] Telfer A. L. and Emmett A. J. (1994), The artificial recharge of stormwater into a dual porosity aquifer, and

the fate of selected pollutants, Proceedings of Water Down Under '94, Adelaide 21-25 November 1994.

[15] Turner J. V., Allison G. B. and Holmes J. W. (1984), The water balance of a small lake using stable isotopes and tritium, Journal of Hydrology, 70, 199-220.

[16] Whipple W. Jr. and Hunter J. V. (1979), Petroleum hydrocarbons in urban runoff, Water Resources Bulletin, 15(4), 1096-1105.

Coupled Modeling of Groundwater and Surface Water for Integrated, Sustainable Water Management in Wetlands

B. Monninkhoff   S.O. Kaden

(WASY GmbH, Institute for Water Resources Planning and
Systems Research, Belin, Germany)

1 Introduction

The main character of wetlands is the frequent availability of water. This can be surface water or groundwater. In most cases, the wetlands are supplied with both types of water. During the raining season surface water is flowing into the wetland after which it is partly infiltrated into the groundwater. During the dry season groundwater is exfiltrated into the wetland from which it is diverted towards the surface water. A special kind of wetland is a polder, at which the wetland is completely diked to prevent the area from flooding with surface water. Both the surface water penetrating the dikes as well as the groundwater which exfiltrates into the diked wetland is usually pumped out of the area. Depending on the amount of water being pumped, the area can still be wet enough to call it a wetland.

Wetlands and polders are difficult to be represented within a detached surface water or groundwater model. The interaction between the groundwater and surface water is both strongly correlated to local surface water and groundwater levels and the resistance between both water bodies. Furthermore, these water levels are sensitive to any changes in the infiltration or exfiltration rates, especially in groundwater dominated wetlands like polders. Most detached surface water and groundwater models offer integrated modules to take this affect into account. These modules either implement simplified basic equations of the flow type which is represented by the module or try to describe the related flow processes with an adapted basic flow equation of the main model. Some models even use a simple volume box to represent the body which is connected. Consequently, each of the mentioned type of modules cannot describe the flow processes within the connected body as accurately as the original detached model. In this paper it is shown that by coupling the two detached models during runtime an effective description of the interaction between the surface water and groundwater bodies can be achieved, even in 2D objects like polders or wetlands. This coupling was successfully tested within a recently finished project in eastern Germany: the Lower Havel area.

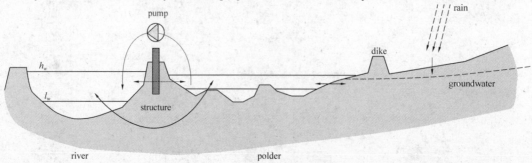

Figure1   principle of a polder

2 Project describtion

The reason for this project was the flood disaster in August 2002 along the river Elbe, one of the most important rivers in northern Europe. This event is classified as the highest ever calendared flood of the river Elbe in the Czech Republic and the federal state of Saxony (Germany). In the federal states downstream of Saxony the flood dimension became more alleviate due to the opening of weirs, the use of the Lower Havel polders as an additional retention area as well as dike failures along the Elbe itself. The controlled flooding of the Lower Havel area during the flood disaster in 2002 clearly showed its particular importance for the reduction of extreme flood events for downstream areas. However, ecological and economical problems within the inundated polder areas made clear that the flooding has its negative side as well.

Starting with a detailed analysis of the Elbe flood in 2002, the management of the weir group Quitzöbel, at which the flood wave in the Elbe is being diverted towards the Havel and its polders, as well as the flooding of the polders itself the project mainly focused on the optimization of the flooding system of the Lower Havel area. As a result a concept for a controlled polder flooding had to be developed. Eventually, this concept has to assist the State Authority for Agriculture and Environment (LUA) Brandenburg and the State Authority for flood protection and water management (LHW) with the development of a new weir operation manual (the present one was finished in 1993 (WBV, 1993)).

For this optimization the software MIKE11® (DHI, 2002) has been used to create a 1D hydrodynamic model, which integrates the Elbe from Tangermünde to Wittenberge, the Havel from Albertsheim to Gnevsdorf, several tributaries to the Havel as well as the 7 Havel polders, each of them divided in a number of sub-catchments.

The Lower Havel area (77 km²) is characterized by close interaction between surface water and groundwater. This has to be considered also with the polder flooding. The groundwater storage offers an additional retention area and is therefore increasing the flood reduction along the downstream part of the Elbe. In former projects this influence could never be accurately quantified (Bronstert, 2004). Depending on the aquifer storativity, the infiltration resistance and the conductivity of the topsoil the share of groundwater storage in the whole polder volume is changing from polder to polder. To analyze the interaction between the groundwater and the flooded polder water a coupled surface water and groundwater model was developed. The groundwater model has been created with FEFLOW® (Diersch, 2002), a software system developed by WASY.

The MIKE11 and the 3D FEFLOW models were coupled using the interface IFMMIKE11 (Monninkhoff, 2004). The interface has been extended to meet the special wishes of this project. The emphasis of this article is not to describe the accomplished model calculations and representation of appropriate results. It rather describes the methical working process for the implementation of the natural conditions.

3 Models

3.1 MIKE 11 Model

In total the model consists of 1,108 MIKE11 branches (including Elbe, Havel, 18 Havel tributaries, 228 sub-catchments and 469 link channels, which connect these sub-catchments), 3,517 HPoints (MIKE11 numerical calculating points) and 57 structures (including all relevant weirs within and along the Havel as well as both the artificial and natural dike breaches towards the polders). Elbe and Havel could be modeled in an arbitrary way with the MIKE11 1D model. The only problem to be solved was to model the weirs Garz and Grütz, which are both spikes barriers. Here the main branch had to be cut in several branches which all represent a different period in which a different number of spikes were set. Due to the width-dependent contraction of these weirs also the resistant values had to be different for each branch.

More complicated was the modeling of the polder flooding with a 1D model. It was chosen to divide each polder in a number of sub-catchments. Each of them is relatively clear distinguishable within the generated 5 m digital terrain model (DTM). They are bordered by dikes, streets or natural high grounds.

Each of the digitized 229 catchments is represented in the MIKE11 model by one single regular branch, which itself only has two profiles. The length of the branch was set to be equal to 50% of the square root of the total area of the sub-catchment. With ArcGIS®, for each of the catchments a water level—area relation could be generated. As each sub-catchment branch in MIKE11 has two cross sections, these areas were divided by two. The profiles in MIKE11 expect a width for each water level. These widths were determined by dividing the area values by half of the length of the branch. These values could be automatically imported in the MIKE11 cross section file.

Each sub-catchment is connected to at least one other sub-catchment. The natural or artificial border which defines both sub-catchments is represented in the model by a so-called link channel. A link-channel is a special MIKE11 branch which doesn't need cross sections and is defined by a table of width-surface level relations. In fact, MIKE11 regards such a channel as a structure, which can be compared with a regular weir. All 469 obligatory width-surface level tables could automatically be extracted from the DTM and afterwards also automatically be imported into the MIKE11 network file.

A large number of these link-channel borders are not completely closed; they are penetrated by culverts or open channels. These crossings are integrated in the model as link-channels as well. As no detailed information about these crossings was available, here a standard cross-section bottom width of 1 m was assumed. Finally, the catchments were connected to the main river (Havel or one of its tributaries) by additional branches. Within these branches a structure represents the dike breach or the weir through which the polder was or can be flooded.

3.2 Interface IFMMIKE11

The coupling of FEFLOW and MIKE11 is not an iterative coupling. That means that no time step is calculated twice, neither by FEFLOW, nor by MIKE11. After each FEFLOW time step, discharges calculated by FEFLOW at the coupled boundary nodes (only 3rd kind, Cauchy type) are exported to the MIKE11 HPoints (calculation points of a MIKE11 network) as an additional boundary condition (Qbase). MIKE11 will calculate its time step as often as needed to reach the actual time level of FEFLOW. If this has been completed, the actual water levels of the MIKE11 HPoints will be exported to the FEFLOW coupled boundary nodes and FEFLOW can start its next time step.

The internal time step of MIKE11 will be controlled by the interface. This time step can be constant or adaptive to the dynamics of the model. The time step of the groundwater model is controlled by FEFLOW. For each of the boundary nodes of FEFLOW, which are selected to be coupled, the closest MIKE11 HPoint is patched directly, which means that the water level of the HPoint is set to the boundary node without considering the chainage and therefore not by interpolation. The same principle is true for the discharges calculated by FEFLOW at the boundary nodes, which are set to the MIKE11 HPoints as an additional boundary. This automatically patching can be additionally controlled using the

FEFLOW feature Observation Point Groups. Here, a single HPoint can be forced to be assigned to a certain group of FEFLOW boundary nodes. Besides Observation Point Groups, the FEFLOW feature Reference Distributions can be used to control the patching as well.

3.3 FEFLOW Model

The model area encloses an area of $77\,km^2$. Only the upper aquifer including locally alluvial sediments is interesting for the model investigations. Therefore the groundwater model consists of 2 model layers, which correspond to these hydrogeological units.

Because of the coupled modeling the mesh generation was subject to very high precision demands. Both in the surface water model and in the groundwater model the location of coupled water courses and inundated areas have to agree accurate, but not exact. Within the working process additional changes and corrections of the FEFLOW mesh were necessary. The FEFLOW supermesh for the mesh generation consists of 72 elements. Altogether the Finite Element Mesh consists of 143,379 nodes and 188,950 elements. The cell size amounts $25 \sim 150\,m$ in the inundated areas. This small size guaranteed that at least several triangles are included in each single sub-catchment within each polder. FEFLOW only calculates fluxes between the surface water and groundwater for boundary nodes which in horizontal direction belong to a triangle which itself has a boundary node at all 3 corners or for boundary nodes which in vertical direction belong to a square which itself has a boundary node at all 4 corners. If one of the corner nodes of a triangle or square doesn't have a boundary the interaction between groundwater and surface water will be neglected for this complete side area of the cell. The smaller the size of the cells the less this phenomenon is influencing the result.

With the interface manager it is possible to extract the calculated fluxes at each single boundary node. In FEFLOW, this flux is calculated backwards out of the balance in this point. During the project it was observed that the flux at one single node is related to the fluxes at the remaining boundary nodes of the side area in which this node is located. This can result in unwanted effects close to dikes. For example, if one of the boundary nodes of a (horizontal) triangle is located at the river side of the Havel dike, both other nodes however are located within the polder at the opposite side of the dike, the fluxes within the polder are influenced by the large flux from the node at the river side of the dike. It was observed that in some cases, the flux inside the polder was directed towards the groundwater although the opposite was to be expected (it normally should be directed towards the surface water in case the polder isn't flooded yet). To minimize the influence of this phenomenon also here the size cell had to be reduced. Beside, the interface was adapted: if the flux direction is opposite towards the expected direction, which can be estimated by the interface knowing both the groundwater as well as the surface water level at the beginning of the time step, the flux is set to zero.

The FEFLOW model was implemented with 3 pre-defined IFMMIKE11 distributions (Ignore, Catch and DTM). Each of the distributions got the correct values using ArcView®. First, the mesh was exported to shape format, which could be imported in ArcView. In ArcView it was possible to give each point of the mesh the correct value of each distribution. The resulting point-file could then be imported to FEFLOW using the 1-neighbour IDW interpolation. In that way each node received the correct value without interpolation. This process was important, because the distributions Ignore and Catch are based on ID's, not on values.

3.4 PROJECT Results

The total model was calibrated for the period August 1 to September 30,2002. The groundwater model calibration shows a good adaptation between measured and computed groundwater dynamics.

The variation between measured and computed groundwater levels amounts predominantly 10 ~ 40 cm. As stated before, the main use for the groundwater model was to estimate the groundwater storage capacity of the aquifer. For this task the reached processing status was sufficient.

Both the Elbe gages as well as the discharge was diverted into the Havel are shown. The differences between measured and calculated values are acceptable. The same is true for the calculated water levels at all Havel gages. To verify the catchment approach satellite images were used to derive the flooded areas for different days. These maps could be compared with the flooded areas calculated by the model. Also here, the results were satisfying.

As stated before, one of the topics of the project was to analyze the influence of the groundwater storage. It was found that during the flooding of 2002 a maximum discharge of ca 25 m³/s was infiltrated into the groundwater. Compared to the maximum inflow into the Havel (more than 700 m³/s) this seems rather irrelevant. However, due to the rather long period of high water levels along the Havel after the flooding (due to the reconstruction of weir Gnevsdorf only a limited discharge was allowed to be diverted to the Elbe to relieve the Havel), this infiltration took place for more than 10 days (the flooding itself took less than 3 days). The balance of the complete Havel catchment between Quitzöbel and Albertsheim is shown. It can be concluded that since the closing of weir Quitzöbel at August 18 at the time of the maximum additional surface water storage (August 27, 141×10^6 m³) ca 9.9×10^6 m³ of the total inflow into the Havel (ca 151×10^6 m³) was infiltrated into the groundwater (ca 9.9×10^6 m³). This amounts about 6.5 %.

Another aspect of the project was to optimize the flooding. This optimization on one hand should reduce the water levels in Wittenberge (which is the downstream end of the model). On the other hand, however, the water level in Havelberg Stadt (Havel) is not allowed to exceed the level of 26.40 m ü NN. It was found that changing the times and locations at which the polders are opened are rather irrelevant in reducing the flood in the Elbe downstream of Wittenberge. Much more important is the time at which the weir Neuwerben is opened and how much water is diverted. The optimization of these two parameters lead to an additional reduction of the maximum water table in Wittenberge of 25 ~ 30 cm (compared to the reduction achieved by the real flooding in 2002). Opening one additional polder (polder 6 has not been flooded in 2002) even 41 cm of additional reduction could have been reached. During the project the flooding system was optimized for different hydrological conditions as well. These scenarios represent reducing a more voluminous wave, a higher wave, a comparable wave but with higher initial water levels in the Havel and combinations of these three.

4 Conclusions and outlook

The project showed that the implemented software IFMMIKE11 is reliable tool to couple the models MIKE11 and FEFLOW. Both the observed groundwater tables as well as the measured surface water levels could be modeled accurately by the coupled system. It was tested that this was not the case using detached models. The project was used to extend the existing version of the coupling software in order to be able to model quasi 2D flow processes in polders. The coupling of regular MIKE11 branches to FEFLOW has already been successfully tested in other projects. So the software is not limited to projects related to wetlands or polders, it could contribute to a more comfortable modeling in any project related to groundwater and surface water interaction.

Future development of IFMMIKE11 will concentrate on an iterative coupling, implement the coupling of mass transport and extend the possibilities to describe the resistance between the surface water and the groundwater bodies. This will be the main topic of the first authors PhD study at the Hohai University of Nanjing

Rerferences

[1] WBV(1993), Vereinbarung zwischen den Ländern Brandenburg und Sachsen-Anhalt über die Bedienung der Wehrgruppe Quitzöbel zur Abwehr von Hochwassergefahren, Potsdam and Magdeburg.

[2] Danish Hydraulic Institute (2002): Reference manual and user manual for MIKE11 River Model, DHI, Copenhagen.

[3] Diersch, H.-J. (2002): FEFLOW Reference manual, WASY GmbH, Berlin.

[4] Monninkhoff B. (2004): Coupling of groundwater model FEFLOW with the hydrodynamic model MIKE 11 (DHI), Tagungsband zur 6. Fachtagung Grafikgestütze Grundwassermodellierung, IWU- Tagungsberichte, 6. Fachtagung: 55-68, Berlin.

[5] Bronstert A (Hrsg.) (2004): Möglichkeiten zur Minderung des Hochwasserrisikos durch Nutzung von Flutpoldern an Havel und Oder. Schlussbericht zum BMBF-Projekt im Rahmen des Vorhabens "Bewirtschaftungsmöglichkeiten im Einzugsgebiet der Havel", Universität Potsdam.

Berlin Water Supply—an Example of Conjunctive Use of Groundwater and Surface Water

Luo Junfeng

(WASY GmbH Institute for Water Resources Planning and Systems Research)

1 Introduction

Berlin, with a population of about 3.4×10^6, is situated in the eastern part of Germany and surrounded by the Bundesland Brandenburg. Water supply in Berlin has been covered primarily by groundwater abstraction from a shallow aquifer consisting of glacial sand and gravel sediments. Waterworks in Berlin are located mostly near rivers or lakes. The use of surface water resources takes place then as:

(1) natural groundwater recharge through bank infiltration.

(2) artificial groundwater recharge by infiltration ponds or basins built near abstraction wells.

Surface water filter through the soil, percolate into and recharge groundwater, become then available for the abstraction. The contribution of surface waters to the total amount of groundwater abstraction is up to 60%~70%. Advantages of using surface water to recharge or replenish groundwater instead of direct using surface water for water production become obvious: during filtration passage in underground mechanical and biogeochemical processes occur. After such natural treatment processes constituents are removed or significantly reduced, that consequently improves the quality of infiltrated waters.

The present paper intends to introduce a 3D groundwater model for a large waterworks in Berlin and its applications for characterisation and analysis of water balance and interactions between groundwater and surface water, subsequently for a sustainable utilisation of the water resources considered.

2 Site information

Berlin's landscape was shaped by glaciers during the last ice age. The City Berlin covers an area of 892 km², of which 55% is built up (urban areas), 35% is covered by green lands or forests and about 7% by surface water. The river Spree flows from SouthEast to NorthWest, crosses through the City Berlin and meets the river Havel, which flows from North to South through western Berlin. The course of the Havel is more like a chain of lakes, the largest being the Tegeler See and Großer Wannsee. A series of lakes also feeds into the upper Spree, which flows through the Großer Müggelsee in eastern Berlin.

The city centre lies along the river Spree in the so-called Berlin-Warsaw glacial valley formed during the ice age about 10,000 years ago. The valley lies between the Barnim plateau in the north and the Teltow plateau in the south. The valley strata consist of sand, gravel, marly till and clay, build valuable shallow aquifers, from which 9 active waterworks, together with a number of smaller industrial or private wells, abstract groundwater for drinking water, industrial or irrigation purposes. All the waterworks of Berlin are located near rivers or lakes.

In 1989, 378×10^6 m³ were abstracted, whereas in 2005 the abstraction dropped to 212×10^6 m³. Due to the overall abstraction of the waterworks for drinking water purposes has been reduced by over 40% during the past 16 years. This mainly was caused by socio-economic changes after German unification.

The Waterworks (WW) Tegel considered in this paper is located in NorthWestern of Berlin and has 7 well galleries around the lake Tegel, The Waterworks Tegel abstracts annually $(45 \sim 60) \times 10^6$ m³ groundwater through 130 vertical wells and one horizontal well. Besides bank infiltration and natural groundwater recharge of precipitation from landside, the WW Tegel uses infiltration ponds and takes surface water from the Lake Tegel (after micro-straining) for controlled artificial groundwater recharge. The annual groundwater recharge ranges from $(9 \sim 19.5) \times 10^6$ m³ in correspondence to the annual abstraction amount. The horizontal well is located on an island in the middle of the Tegeler See, surrounded by other well galleries and has normally an average pumping rate of 5,000 m³/d.

In the surrounding of the Waterworks Tegel there are another three large waterworks which abstract groundwater from the same aquifer with an annual amount of 88×10^6 m³.

Due to a complex hydrogeological situation (see following chapter) and large amounts of groundwater pumping as well as artificial groundwater recharging, a complicated groundwater flow field and dynamics occur in the surroundings of the Waterworks Tegel. The fluctuation of groundwater level there can reach about 8 m. In a short distance of 5 m near to the horizontal filter pipe of the horizontal well, a large water level difference has been observed to 4 m.

In the surroundings of the Waterworks Tegel there are several known contaminated sites endangering sustainable utilisation of the water resources considered.

3 Hydrogeological situation

The aquifer system considered can be characterised by a hydrogeologic section. According to that there are three main aquifers of glacial sediments in the working area. The upper aquifer (1st aquifer: GWLK 1) is formed by glaciofluviatile fine to coarse sand with a thickness of about 10 m and occurs phreatic or unconfined. The 2nd one (GWLK 2), also called main aquifer of quaternary Saale ice-age, where abstraction wells are screened, consist of about 30 m thick fine to coarse sand and partly gravel and is predominantly covered by glacial till of about 4 m thickness. Since the aquitard of till is locally absent, the 1st and 2nd aquifer are hydraulically connected. The 3rd (GWLK 3), a tertiary fine sand aquifer with 70 m thickness, lies below and separated from the two upper aquifers by $20 \sim 30$ m thick glaciofluvial silt and clay of the Holstein interglacial or Pleistocene.

The Lake Tegel itself is formed during the quaternary Saalian ice-age and actually has a maximum depth of 14 m, with the deeper regions already being filled up with thick silt sediments or mud, decreasing and vanishing towards the lake bank. The silt or mud at the lake bottom reduces or obstacles interactions between surface water and groundwater.

4 Groundwater model

In a contract of the Berlin Water (Berliner Wasser Betriebe, BWB) a three dimensional regional groundwater model has been built with Simulation software FEFLOW®. The groundwater model consists of 10 numerical Layers taking following hydrogeologic or hydraulic units into consideration:

Unit 1: Covering layer-Peat, mud or fluvial sand of Holocene.

Unit 2: 1st Aquifer (GWLK 1) of Weichselian-Eemian ice age.
Unit 3: Aquitard 1/2 (GWH 1/2) of Warthian ice age.
Unit 4: 2nd Aquifer (GWLK 2) of Saalian ice age.
Unit 5: Aquitard 2/3 (GWH 2/3) of Holsteinian Interglazial.
Unit 6: 3rd Aquifer (GWLK 3) of Elsterian ice age and tertiary.

The implementation of the hydrogeological model gives an overview of the built 3D-Groundwater model with 266,510 finite elements and 149,864 nodes.

The model area with about 270 km² encloses the four big Waterworks with 258 pumping wells, river Spree and river Havel, ditches, channels and lakes. Specially, the horizontal well was implemented through the functional option of "Discrete feature elements" provided by FEFLOW®.

5 Calibration of the groundwater model

The groundwater model has been calibrated for both steady and transient flow of the year 2002, i.e.:

(1)Simulation of average groundwater flow filed in 2002 with input of annual means values of groundwater abstraction, artificial groundwater recharge, surface water tables and natural groundwater recharge of precipitation in 2002.

(2)Simulation of groundwater flow dynamics during 2002 with input of monthly means values of groundwater abstraction, artificial groundwater recharge, surface water tables and natural groundwater recharge of precipitation in 2002.

The simulated values have been compared with measured groundwater levels in 366 observation wells: 217 in the 1st Aquifer (GWLK 1) and 149 in the 2nd Aquifer (GWLK 2). Results show that both the regional groundwater flow and local dynamics in the horizontal well have been successfully reproduced by the FEFLOW model.

6 Applications of the groundwater model

The calibrated groundwater model can be then applied to analyse of water balance and interactions between groundwater and surface water under consideration of different cases water resources planning and subsequently for sustainable utilisation of the water resources considered.

6.1 Water balance analysis for the situation of 2002

The annual precipitation of 2002 in the model area amounts to 733 mm, from which 166 mm as natural groundwater recharge are available in the total model area. The annual abstraction amount of the Waterworks Tegel is about 48.0×10^6 m³. About 11.2×10^6 m³ surface water have been infiltrated into underground to replenish groundwater and contributed to 23.3% of the abstraction amount. The calculated annual bank infiltration is about 27.1×10^6 m³ and 56.3% to the abstraction amount. The surface water contributes then about 79.6% to the total abstraction amount. The annual natural groundwater recharge consequently contributes only 20.3% to total abstraction amount.

6.2 Planning Case study for year 2004

The planning case for the year 2004 has been investigated under the following considerations:

(1)Annual average natural groundwater recharge of 112 mm estimated from precipitations of long-term series 1976~2002 (annual average: 538 mm).

(2)Annual abstraction amount of the Waterworks Tegel is 45.0×10^6 m³ and its monthly distribution or partition to certain abstraction wells is given, modifications only in case of needed.

(3)Annual artificial groundwater recharge 7×10^6 m³ available and its monthly distribution to infiltration ponds is given.

(4)Controlling and monitoring groundwater levels in urban and industrial areas south to Waterworks Tegel.

(5)Controlling and monitoring potential changes of groundwater flow fields due to known contaminated sites around the Waterworks Tegel.

In the following main results are briefly described. More details of model development and application will be presented online during the workshop.

The estimated contribution (incl. the amount of artificial groundwater recharge) of surface water to the abstraction amount is about 80.6%.

It simulated groundwater levels (lilac lines) according to planning case 2004 in comparison with the groundwater levels which have to be kept as possible (blue isolines). It indicate that the simulated groundwater levels in areas south to the Waterworks Tegel are 0.25 ~ 0.75 m higher as it should be.

Other shows calculated forward pathlines from known or suspected contaminated sites, indicating potential changes of groundwater flow fields in comparison with status quo.

It presents backward pathlines and catchments of well galleries with given monthly abstractions according to planning case 2004.

Nitrogen Transport in Soils Under the Condition of Sewage Irrigation

Yang Jinzhong   Wang Liying

(National Key Laboratory of Water Resources and Hydropower Engineering Sciences,
Wuhan University, Wuhan, China)

1 Introduction

The FILTER (Filtration and Irrigated cropping for Land Treatment and Effluent Reuse) technique was developed by Jayawardane (1995) to address problems of effluent irrigation on soils with restricted internal drainage, where salinity and soil aeration problems could develop. The FILTER technique combines nutrient and water reuse by applying the nutrient-rich effluent to grow crops with filtration through the soil to a sub-surface drainage system. The FILTER system is thus capable of handling high volumes of effluent during periods of low cropping activity or high rainfall, allowing wastewater treatment throughout the year and eliminating the need for wastewater storage. In this system the rate of effluent application and drainage could be regulated to ensure the required level of pollutant removal, thereby producing low-pollutant drainage water suitable for discharge to surface water bodies.

Modelling of the FILTER system can be used to increase the operation efficiency of the FILTER system in removing pollutants and to predict any long-term adverse effects on the surrounding environments. This paper describes the development of a Nitrogen-2D model to simulate the nitrogen dynamics during wastewater application to FILTER systems. The wastewater flow is described using the 2D USDA SWMS-2D model (J Simunek et al., 1994). Nitrogen movement in soil and groundwater is driven by water flow and affected by temperature, pH and oxygen. The basic equations describing these processes are combined with transformations of nitrogen, such as mineralization, root uptake, nitrification, volatilization and adsorption. The model predictions of nitrogen movement and transformations are tested against field data from experimental FILTER plots and that from lysimeters in China.

2 Mathematic model

2.1 Nitrogen transformation and root uptake

2.1.1 Nitrogen pools

There are two pools for the organic matter. One is the Litter pool, which is composed of crop residual, dead root, and microbial biomass. Another is Humus pool, which is composed of the stabilized decomposition products. The flow of carbon and nitrogen in the organic pools is shown in Figure 1.

q_c is the carbon change rate because of the decomposition of the organic matter in Litter pool. f_e is the synthesis efficiency. f_h is the humification fraction.

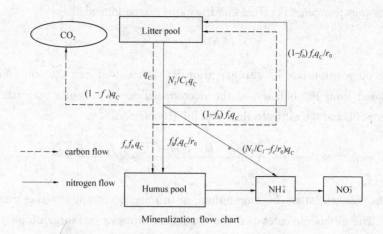

Figure 1 Carbon and nitrogen flow in organic pools

2.1.2 Mineralization and immobilization

Like many other nitrogen models, such as SOILN, ANIMO and DAISY (L.Wu et al.,1998), Nirogen-2D represents a transformation or flow between pools in most instances as a "first-order rate process", which means that the flow out of the first pool to the second is proportional to the quantity of material remaining in the first pool. The decomposition rate of carbon in Litter pool is expressed as first-order kinetics.

$$\left. \begin{array}{l} \dfrac{\partial C_l}{\partial t} = -K_l f_l C_l \\ f_l = f_\theta f_T f_{C/N} f_{pH} \\ C_l = C_{l,i} \\ t = t_i \end{array} \right\} \quad (1)$$

where, C_l is carbon content in Litter pool. K_l is the rate coefficient. f_θ, f_T, $f_{C/N}$ and f_{pH} are the response function of water content, temperature, C/N ratio, and pH respectively. $C_{l,i}$ the carbon content at time t_i.

The decomposition rate of nitrogen in Litter pool is calculated from the C/N ratio and the decomposition rate of carbon.

$$\dfrac{\partial N_l}{\partial t} = \dfrac{1}{(C/N)_l}\dfrac{\partial C_l}{\partial t} = \dfrac{N_l}{C_l}\dfrac{\partial C_l}{\partial t} \quad (2)$$

where, $(C/N)_l$ is the current C/N ratio in Litter pool. N_l is the nitrogen content in Litter pool.

Since there is one pool to represent the litter-decomposer pool, the synthesis of biomass and metabolites constitutes an internal cycling of carbon return to the Litter pool. The returned rate of carbon content is

$$(RC_l)_{l\to l} = -(1-f_h)f_e \dfrac{\partial C_l}{\partial t} \quad (3)$$

The model assumes that the products of synthesis from biomass and metabolite have the same C/N ratio r_0 as the humus pool. With the internal carbon flow in Litter pool, the nitrogen content change rate from the internal circling is

$$(RN_l)_{l\to l} = -\dfrac{(1-f_h)f_e}{r_0}\dfrac{\partial C_l}{\partial t} \quad (4)$$

The nitrogen content change rate in Humus pool because of carbon flow from Litter pool into

Humus pool (this transformation is called humification) can be indicated as

$$(RN_h)_{l \to h} = -\frac{f_h f_e}{r_0}\frac{\partial C_l}{\partial t} \tag{5}$$

The net decomposition rate of nitrogen from the Litter pool into the soil solution (in inorganic form) is calculated from the balance of the decomposition of nitrogen in Litter pool, the internal circling in Litter pool, and the nitrogen flow into the Humus pool.

$$\frac{\partial N}{\partial t} = -\left[\frac{N_l}{C_l}\frac{\partial C_l}{\partial t} - \frac{(1-f_h)f_e}{r_0}\frac{\partial C_l}{\partial t} - \frac{f_h f_e}{r_0}\frac{\partial C_l}{\partial t}\right] = -\left[\frac{N_l}{C_l} - \frac{f_e}{r_0}\right]\frac{\partial C_l}{\partial t} \tag{6}$$

where, N is the concentration of ammonium or nitrate according to the mineralization or the immobilization. The minus indicates that the change of nitrogen in inorganic pool is same as that in organic pool, but with an opposite direction.

The mineralization of the Humus pool is calculated by a first-order kinetic when finish the calculation of receiving nitrogen from Litter pool.

$$\left.\begin{array}{l}\dfrac{\partial N_h}{\partial t} = -K_h f_h N_h \\ f_h = f_\theta f_T f_{C/N} f_{pH}\end{array}\right\} \tag{7}$$

where, N_h is nitrogen content in Humus pool. K_h is decomposition rate coefficient of Humus pool.

The general forms of transformation rates of nitrification, denitrification, and volatilization.

(1) zero-order kinetics:

$$\frac{\partial \theta N}{\partial t} = -K_0(\theta + \rho K_d) f_\theta f_T f_{C/N} f_{pH} \tag{8}$$

(2) first-order kinetics:

$$\frac{\partial \theta N}{\partial t} = -K_1(\theta + \rho K_d) f_\theta f_T f_{C/N} f_{pH} N \tag{9}$$

(3) Michchaelis-Menten kinetics:

$$\frac{\partial \theta N}{\partial t} = -K_m f_\theta f_T f_{C/N} f_{pH}\frac{(\theta + \rho K_d)N}{(\theta + \rho K_d)N + K_C} \tag{10}$$

where, N is the ammonium or nitrate concentration in soil solution. K_0 is the zero-order rate constant; K_1 is the first-order rate constant. K_d is the distribution coefficient for adsorbing solute. K_m is the maximum rate constant. K_C is the saturation coefficient. f_θ, f_T, $f_{C/N}$ and f_{pH} are the response functions from the water content, temperature, C/N ratio, and pH in soil.

All the nitrification, denitrification, volatilisation and root uptake can choose one of above form of kinetics depending on the measurement data and environment of the simulation domain. $R_{\theta N} = \partial \theta N / \partial t$ is the change rate of nitrogen concentration, which depends on the kinetics of transformation and is added to the ammonium or nitrate transport equations as a source terms. $R_{\theta N}$ can be used to express the denitrification rate R_{de}, nitrification rate R_n, and volatilisation rate R_v.

2.1.3 Root uptake

Two models can be chosen to denote the nitrogen uptake by plants.

2.1.3.1 Model 1

The potential root uptake rate of nitrogen is expressed as (Johnsson et al., 1987)

$$RN(t) = \frac{\partial UN(t)}{\partial t} = UN(t)a\left(1 - \frac{UN(t)}{UN_a}\right) \tag{11}$$

where, $UN(t)$ is the potential cumulative nitrogen in the plant at the time t from the start of growing season, which can be expressed as a logistic function

$$UN(t) = \frac{UN_a}{1 + \frac{UN_a - b}{b} e^{-at}} \qquad (12)$$

where, UN_a is the annual potential nitrogen uptake, a and b are experimental parameters.

The root uptake rate of nitrate has the following form

$$R_{p3} = \frac{\partial N_3}{\partial t} = \min\left(\beta(x,z)\frac{N_3}{N_3 + N_4}RN(t) \times L_s, f_{NM}N_3\right) \qquad (13)$$

where, f_{NM} is maximum available nitrogen fraction in soil solution. L_s is the ground surface length. $\beta(x,z)$ is the normalized root distribution modified from APEX (Williams.J.R. et al.,2005), which can be estimated with the function,

$$\begin{aligned}\beta(x,z) &= \frac{\left[\frac{5.0 \times RZ}{RD}\exp(\frac{-5.0 \times RZ \times z}{RD})\right]}{[1.0 - \exp(-5.0 \times RZ)]} \\ RD &= \min(2.5 \times RD_{max} \times HUI, RD_{max}, RZ)\end{aligned} \qquad (14)$$

where, RD is the root depth in m. RD_{max} is the maximum root depth in m for crop. HUI is heat unit index ranging from 0 at planting to 1.0 at physiological maturity is computed by accumulating daily HU (the number of heat units accumulated during a day) values and dividing by the potential heat units of the crop. Function (14) assumes that the crop is uniformly distributed on the ground surface.

The uptake of ammonium is estimated with the same function as equation (13)

$$R_{p4} = \frac{\partial N_4}{\partial t} = \min\left(\beta(x,z)\frac{N_4}{N_3 + N_4}RN(t) \times L_s, f_{NM}N_4\right) \qquad (15)$$

2.1.3.2 Model 2

The root uptake rate of nitrogen can also be simply treated as being proportional to the root uptake rate of water and the concentration in bulk soil solution (Rijtema and Krose, 1991; Zhang et al., 1997), that is

$$R_{p4} = C_4 f_R S \quad R_{p3} = C_3 f_R S \qquad (16)$$

where, C_4 and C_3 are the ammonium and nitrate concentration in soil solution. f_R is the root uptake coefficient ranging from 1.0 to 1.25. S is the root uptake rate of water.

2.2 Water flow and Nitrogen transport model

2.2.1 Water flow in saturated and unsaturated soils

Nitrogen-2D model is designed to solve 2D saturated-unsaturated flow equations under specified initial and boundary conditions, relevant to the FILTER system.

The water flow equation is expressed as

$$\frac{\partial \theta}{\partial t} = \frac{\partial}{\partial x_i}\left(K_{ij}\frac{\partial h}{\partial x_j}\right) - \frac{\partial K_{i2}}{\partial x_i} - S \qquad (17)$$

Where, θ is volumetric water content. h is the pressure head. S is the sink(source) term. x_i is the spatial coordinates with x_1 in the horizontal and x_2 in the vertical directions. K_{ij} is the components of the unsaturated soil conductivity tensor.

$$K_{ij} = K_r K_s K_{ij}^A \tag{18}$$

Where, K_r is the relative hydraulic conductivity. K_s is the saturated hydraulic conductivity. K_{ij}^A is the dimensionless anisotropic tensor.

The soil water characteristics and unsaturated relative hydraulic conductivity are expressed by van Genuchten model.

When the sink term S represents the root uptake, it is estimated with the function (Feddes et al., 1978)

$$S = \alpha(h) S_p \tag{19}$$

where, $\alpha(h)$ is a pressure response function of root uptake. S_p is the potential water uptake by plant, which is related to the potential transpiration rate.

2.2.2 Nitrogen transport equations

Transport of ammonium and nitrate in saturated and unsaturated soil is expressed by convection-dispersion equation. The adsorption of ammonium on the bulk soil is considered, therefore, the ammonium transport equation is

$$\frac{\partial \theta C_4}{\partial t} + \frac{\partial \rho S_4}{\partial t} = \frac{\partial}{\partial x_i}\left(\theta D_{ij}\frac{\partial C_4}{\partial x_j}\right) - \frac{\partial}{\partial x_i}(q_i C_4) + R_4(C_l, C_h, C_4, C_3) \tag{20}$$

where, C_4 is the ammonium solution concentration. S_4 is the adsorbed concentration on soil particles. ρ is soil bulk density. q_i is Darcy flux in x_i direction. R_4 is the source and sink term, which can be expressed as

$$R_4(C_l, C_h, C_4, C_3) = R_{m4}(C_l, C_h, C_4, C_3) + R_n(C_4) + R_{p4}(C_4) + R_v(C_4) \tag{21}$$

where, R_{m4} is the net mineralization(immobilization) of ammonium. C_l and C_h are the nitrogen content in Litter pool and Humus pool. R_n is the nitrification rate. R_{p4} is root uptake rate of ammonium. R_v is the ammonium volatilization rate. All the source and sink terms can be a non-linear function of the ammonium concentration. Therefore, equation (20) may be a non-linear partial differential equation.

Ion-exchange process governing NH$_4$-N adsorption-desorption is assumed to be of the linear Freundlich form (Carbon et al., 1991)

$$S_4 = K_d C_4 \tag{22}$$

Where, S_4 is adsorbed ammonium concentration on the soil particles. C_4 is the ammonium concentration in soil solution. K_d is distribution coefficient, representing the ratio between NH$_4$-N adsorbed and NH$_4$-N in the soil solution.

Substituting equation (22) into equation (20) and considering the conservative equation of water, we have

$$\frac{\partial N}{\partial t} = -\left[\frac{N_l}{C_l}\frac{\partial C_l}{\partial t} - \frac{(1-f_h)f_e}{r_0}\frac{\partial C_l}{\partial t} - \frac{f_h f_e}{r_0}\frac{\partial C_l}{\partial t}\right] = -\left[\frac{N_l}{C_l} - \frac{f_e}{r_0}\right]\frac{\partial C_l}{\partial t} \tag{23}$$

Disregarding the adsorption of nitrate on the bulk soil, convection-dispersion equation for nitrate transport is

$$\frac{\partial \theta C_3}{\partial t} = \frac{\partial}{\partial x_i}\left(\theta D_{ij}\frac{\partial C_3}{\partial x_j}\right) - \frac{\partial}{\partial x_i}(q_i C_3) + R_3(C_l, C_h, C_4, C_3) \tag{24}$$

The source and sink term R_3 in equation (24) can be expressed as

$$R_3(C_l, C_h, C_4, C_3) = R_{m3}(C_l, C_h, C_4, C_3) - R_n(C_4) + R_{dn}(C_3) + R_{p3}(C_3) \tag{25}$$

where, R_{m3} is the net immobilization from nitrate. R_{dn} is the denitrification rate. R_{p3} is the root uptake rate of nitrate. Following the same way of the derivation of equation (23), we have

$$\frac{\partial}{\partial x_i}\left(\theta D_{ij}\frac{\partial C_3}{\partial x_j}\right) - q_i\frac{\partial C_3}{\partial x_i} - \theta\frac{\partial C_3}{\partial t} = -SC_3 - R_3(C_l, C_h, C_4, C_3) \tag{26}$$

In the above equations, D_{ij} is component of the dispersion coefficient tensor which combined the diffusion of nitrogen in soil and the hydrodynamic dispersion resulting from the variation of pore water velocity, and

$$\theta D_{ij} = \alpha_T q \sigma_{ij} + (\alpha_L - \alpha_T)\frac{q_i q_j}{q} + \theta D_d \tau \sigma_{ij} \tag{27}$$

where, α_L and α_T are longitudinal and transverse dispersivity, respectively. q is the magnitude of Darcy flux. $q = \sqrt{q_i^2 + q_j^2}$, σ_{ij} is the Kronecker delta tensor. D_d is the diffusion coefficient of ammonium or nitrate in free solution.

2.2.3 Temperature calculations

The temperature is calculated from an empirical equation (Rijtema and Kroes, 1991).

$$T(z,t) = T_a + A_0 \exp(-z/D_m)\cos(\omega t + \phi - z/D_m) \tag{28}$$

where, T is the temperature, ℃. T_a is the average yearly temperature, ℃. A_0 is amplitude of temperature wave. D_m is damping depth, m. ω is frequency of temperature wave. ϕ is phase shift; $Z=T(x,t)$.

3 Experiments

One of the field experiments was conducted at the Griffith City, Australia. The soil at the site is a transitional red-brown earth (Stace et al. 1968) or a Typic Chromexert in the Soil Taxonomy (Soil Survey Staff 1975). The site has a high saline groundwater table of about 1.5 m and was previously used for land application of saline sewage effluent for irrigation during summer months, without provision of sub-surface drainage.

The FILTER trials were conducted on eight plots(1 hm^2), starting in summer 1994(1995). Two experimental blocks (A and B), each consisting of 4 FILTER plots were established within a 12 hm^2 area, which was initially laser levelled to a 1 : 4,000 slope. Each plot is 40 m wide by 250 m long, and is surrounded by a 0.4 m high bank. Experimental Block A consisting of plots 5, 6, 7 and 8 was used for summer filtration during the period November 1994 to May 1995. Experimental Block B consisting of plots 1, 2, 3 and 4 was used for winter filtration during the period from May 1995 to November 1995.

The other experiment was carried out during November 2003 to June 2004 at the irrigation and drainage experimental station (Wuhan University), China. The soil at the site is clay loam with a bulk density of 1.42 g/cm^3. The site has a fixed groundwater table of 2.9 m that can be adjusted to meet the needs of different experiments. Irrigation experiment was conducted on four lysimeters when winter wheat was growing, three of which (number 9, 10 and 18) were irrigated with sewage effluent, and fresh water was applied on lysimeter number 8. Each lysimeter has the dimension of 2×2×3 m^3, where the water content, soil temperature, soil water electric conductivity were measured. The lysimeters are covered with a rain protection sheet 4 m above the ground, which can hold up the rainfall if necessary. The experiment lasted 198 days from sowing to harvest, and during this time, rain were hold up from the lysimeters, and the winter wheat was irrigated three times with a cumulative volume of 306 mm sewage effluent or fresh water. The sewage effluent has an average inorganic nitrogen concentration of 30 mg/L.

4 Numerical Simulations

The field experimental block A is used to calibrate the parameters needed for the model. These parameters include hydraulic parameters, root uptake parameters, solute transport parameters, nitrogen mineralizing(immobilized) parameters, and nitrifying(denitrifying) parameters. All the parameters calibrated in the field experiment were then used for the simulation of the experiment from block B and the simulation analysis of the experimental data from the lysimeters in the experimental station of China.

4.1 The model calibration

The FILTER experiment in block A was conducted from November 1994 to May 1995, which included 12 effluent irrigating events and lasted for 177 days. The mean volume and nitrogen concentration of the effluent applied in four plots is shown in Table 1. The precipitation was measured at a weather station near the experimental site, and the total rainfall during the 177 days was 172 mm. The inflow to the experimental plots is the sum of the effluent applied to the field and the precipitation, and the concentration of NH_4-N and NO_3-N in inflow is the mean values in the effluent applied and the precipitation. The averaged experimental data from the four plots were used for the calibration.

The soil hydraulic parameters were initially obtained through inverse modelling of water flows, using the HYDRUS-2D model (J. Simunek et al.,1999). The parameters for different layers were further refined using inverse modeling with field data from Block to yield the parameters shown in Table 2. The soil hydraulic parameters in the lysimeters were initially cited from HYDRUS-2D model by soil type and were adjusted according to hydraulic data of number 8 lysimeter, which are also given in Table 2.

The first-order kinetics for nitrification, denitrification and the model 2 for root uptake are used in the modeling. All other parameters related to the nitrogen transformation are cited from literatures and are shown in Table 3.

Table 1 volume and nitrogen concentration of the effluent applied in Block A

Event	Nitrogen concentration of the effluent (mg/L)			Volume (mm)
	NH_4-N	NO_3-N	Organic nitrogen	
1	0.46	0.16	7.30	96.6
2	0	0	7.32	77.0
3	0.23	0	11.47	88.3
4	0.44	0	3.87	194.3
5	0	0	6.26	100.2
6	0.01	0	6.18	115.5
7	0.2	0	2.30	115.5
8	0	0	6.87	115.5
9	0.01	0	7.64	115.5
10	0.03	0	17.07	123.8
11	0.01	0	10.25	115.5
12	0	0	7.67	74.3

Table 2 The soil parameter in van Genuchten model

Depth(m)	0~0.3	0.3~0.6	0.6~0.9	0.9~1.2	Soil near the drainpipes	Soil in lysimeters
θ_r	0.1	0.1	0.1	0.1	0.1	0.095
θ_a	0.1	0.1	0.1	0.1	0.1	0.095
θ_s	0.50	0.48	0.46	0.44	0.50	0.41
θ_m	0.50	0.48	0.46	0.44	0.50	0.41
θ_k	0.49	0.47	0.45	0.43	0.49	0.41
a(m)	10.0	7.5	7.5	5.0	10.0	1.9
n	1.15	1.15	1.15	1.15	1.15	1.31
K_s(m/d)	0.700	0.45	0.30	0.022	0.45	0.524
K_k(m/d)	0.680	0.44	0.29	0.021	0.44	0.524
Bulk density (t/m^3)	1.3	1.3	1.5	1.6		1.42

Table 3 Parameters used in the modeling

Parameter	Meaning	Unit	Value	Source of data
K_l	Rate coefficient for mineralization in Litter pool	1/d	0.003,5	Andren O et al., 1987
f_e	Synthesis efficiency constant		0.5	Kjoller A et al., 1982
f_h	Humification fraction		0.2	Johnsson H et al., 1987
k_h	Rate coefficient of mineralization in humus pool	1/d	0.000,01	Vinten A J A et al., 1996
K_{den}	First-order rate coefficient of denitrification	1/d	0.05	Calibrated
K_{nit}	First-order rate coefficient of nitrification	1/d	0.05	Calibrated
f_R	Root uptake coefficient		1.0	Rijtema and Krose, 1991; Zhang et al., 1997

Because of the symmetry of the experimental site, the simulation domain is selected as shown in Figure 2. The measured daily rainfall and potential evapotranspiration for the whole simulation period were used as a time-variable boundary at the soil surface. The amount of water irrigated was added to the rainfall data. The left, right, and bottom boundaries were assumed to be no-flow boundaries. The drainpipe was simplified to a seepage face. The solute transport boundary is determined based on the direction of water flux on the boundary.

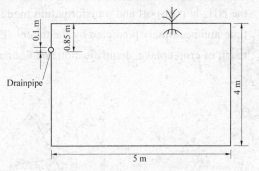

Figure 2 Simulation domain setup

4.2 Simulation result and compared with experimental data

4.2.1 Simulation result of the water table depth in the field experiment

In the irrigation period of the effluent application stage, the water ponded on the soil surface because the cumulated effluent application is larger than soil infiltration. The boundary condition in the effluent application stage can be indicated as

$$\left.\begin{array}{l} \dfrac{\partial h}{\partial t} = I(t) - k_{ij}(\dfrac{\partial h}{\partial x_j} + \sigma_{j2})n_i \quad (h>0, (x_1,x_2)\in R_I) \\ 0 = I(t) - k_{ij}(\dfrac{\partial h}{\partial x_j} + \sigma_{j2})n_i \quad (h\leqslant 0, (x_1,x_2)\in R_I) \end{array}\right\} \quad (29)$$

where, n_i is the unit vector out of the boundary. $I(t)$ is the irrigation rate.

Figure 3 shows the water table depth at the middle point of the parallel spacing drains in FILTER experiment. It is shown that effluent application raises the groundwater table rapidly at the beginning of irrigation, and then the plot is pounded after the soil is saturated. The agreement of the experimental and the simulated results shows the parameters used in the simulation is reliable and the model set-up is reasonable.

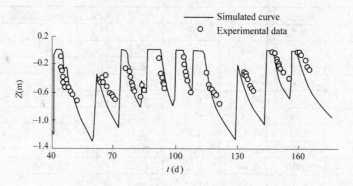

Figure 3 Comparison of the simulated water table depths with the measured data in Block A

4.2.2 Nitrogen concentration and the mass balance in the field experiment

The simulation results of NH_4-N and NO_3-N, together with experimental data of the drainage water from the drainpipe in FILTER experiment are shown in Figure 4. Generally speaking, the simulated results are in good agreement with the experimental data especially for NO_3-N data. The simulated NH_4-N concentration does not match the fluctuation of the experimental data. One of the reasons may be that NH_4-N concentration in the soil is strongly affected by the environmental conditions, which are not considered in the model. Another reason may be partly due to the difficulties in accurately measuring the very low values of NH_4-N in the drainage waters. Further research work should be considered to improve the NH_4-N transport and transformation model. The nitrate concentration of the soil water decreases with time and accurately predicted by the model. The consecutive reduction of soil NO_3-N concentration is the result of crop uptake, denitrification and drainage through the pipe drain system.

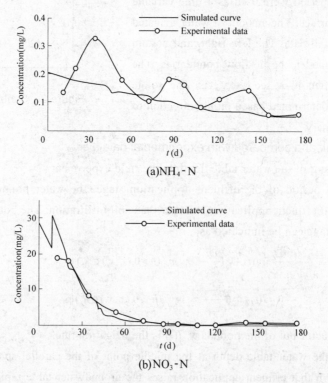

Figure 4 NH_4-N and NO_3-N concentration in drainage outflow in Block A

Table 4 shows the solute balance analysis simulation. The relative error is small and acceptable. The measured solute mass is comparable to the simulated values.

Table 4 Mass balance analysis in the simulation of Block A

Balance terms	NH_4^+ (g)		NO_3^- (g)	
	Model input	simulated	Model input	simulated
CumCh0		1.36		−18.4
CumCh1		−24.2		−139
CumChR		−0.206		−29.8
ChemS1		−0.182		−3.5
ChemS2	0.924	0.924	0.072,2	0.072,2
$M_0 - M_t$		22.9		195.163
sum		0.596		4.54
Abs–Error		0.596		4.54
Rela–Error=(Abs–Error/(M_t−M_0))		2.6%		2.3%

The terms in table 4 are as follows:
  CumCh0—solute removed from the flow region by zero-order reactions (negative when removed from the system).
  CumCh1—solute removed from the flow region by first-order reactions.
  CumChR—solute removed from the flow region by root water uptakes.
  ChemS1—solute flux across drainpipe.
  ChemS2—solute flux across the atmospheric boundary.
  M_0–M_t—change of the solute in the flow region in the experimental period.
  Abs–Error—absolute error.
  Rela–Error—relative error.

It indicates from the above discussion that the comparison of the simulated result with the experimental data is acceptable. This means that the determined parameters are reasonable, and can be used for the prediction of the water movement and solute transport under similar conditions. Therefore, the proposed model is used to predict the ammonium and nitrate distribution under irrigation and drainage condition conducted in Block B.

The experiment in Block B lasted for 160 days with total 12 effluent application events. The other experimental conditions are similar to Block A. The volume and the corresponding concentration of effluent applied is shown in Table 5.

Table 5 volume and nitrogen concentration of the effluent applied in Block B

Event	Nitrogen concentration of the effluent (mg/L)			Volume (mm)
	NH_4-N	NO_3-N	Organic nitrogen	
1	1.58	2.23	8.25	103.0
2	2.66	1.41	9.35	80.0
3	4.46	1.40	11.09	86.5
4	5.92	0.83	8.78	60.0
5	5.52	0.30	9.08	75.0
6	12.36	0	9.12	80.0
7	9.99	0.86	4.31	100.0
8	10.90	0.51	0.1	90.0
9	8.04	0.48	5.06	100.0
10	9.43	0.29	3.51	82.5
11	7.80	0.25	7.60	90.0
12	6.18	0.42	0	100.0

The simulated water table depth is shown in Figure 5, together with the experimental data measured during the drainage period. The simulated water table agrees well with the experimental data, which indicates that the water flow model and the parameters are reliable and can be used in these complicated conditions.

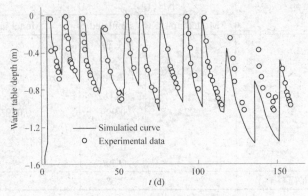

Figure 5 Comparison of predicted groundwater depth with the measured data in Block B

As the simulation result in Block A, the NH_4-N concentration from the measurement has a large fluctuation and the predicted result can just captured the general tendency. The simulation result of NO_3-N is better than that of NH_4-N.

4.2.3 Simulation and analyses for sewage irrigation in lysimeters

The simulation begins on March 8, 2004, and ends on June 1, 2004, and during this time, the lysimeters was irrigated two times with a cumulative volume of 210 mm sewage effluent or fresh water.

Because of the symmetry of the lysimeters, the simulation domain is selected as a rectangle with the dimension of 2.0 m × 3.0 m. The potential evapotranspiration for the whole simulation period were used as a time-variable boundary at the soil surface. The left and right boundaries were assumed to be no-flow boundaries, and the bottom boundary was assumed to have a constant water head of 0.1 m according to the experimental condition. The solute transport boundary is determined based on the direction of water flux on the boundary.

The measured and simulated nitrogen concentration of soil profile on 1st, in June. A good agreement of the experiment data and the estimated value is found except number 8 lysimeter. In number 8 lysimeter, the model has over estimated the nitrate concentration possibly because the nitrogen transformation parameters were cited from the simulation of the field sewage irrigation experiment and they don't fit fresh water irrigation well. Another reason may come from the ignored effects in the model on plant growth between sewage irrigation and fresh water irrigation.

The above results indicate that the model can offer a good prediction tool of soil water content and soil nitrogen content. We can use it to analyse what will happen if the sewage effluent used for irrigation has a higher nitrogen concentration.

4.2.4 Model application

It is shown in the field and lysimeter experiments that the nitrogen concentration and the amount of sewage effluent used in the experiment didn't cumulate in the soil and the drainage water concentration is low. What will happen if the sewage effluent has more nitrogen in it? To answer this question, we design a simulation experiment as following: with the same climate, soil and crop conditions as in the lysimeter experiment, the lysimeter is supposed to be irrigated three times from sowing (November, 5) to harvest (May, 21) with a total volume of 300 mm. The concentration of nitrogen in the sewage effluent has four levels, arranging from 0, 50, 66 mg/L and 83 mg/L ammonia respectively. The results of the simulation are shown in Table 5.

It is shown in Table 6 that about 49% percent (=[nitrification (level)–nitrification(blank)]/application) of ammonia was turned into nitrate by nitrification, 25.3% percent (=[denitrification (level)–denitrification (blank)]/[nitrification (level)–nitrification(blank)]) of the resultant from nitrification loss from soil by denitrification, 46.7% percent (=[root uptake (level)–root uptake (blank)]/[nitrification(level)– nitrification (blank)]) for plant uptake and the rest (28% percent (=100%–25.3%–46.7%)) remains in soil within 1.5 m depth.

Table 6 Nitrogen balance within 1.5 m depth

Application		Change	Root uptake	Nitrification	Denitrification	Mineralisation
Blank (N 0g /m^2)	NH$_4$-N	−11.497	0.186	21.41		10.12
	NO$_3$-N	−3.772	10.97		14.86	
Level 1 (N 15g /m^2)	NH$_4$-N	−4.015	0.393	28.89		10.12
	NO$_3$-N	−1.863	14.46		16.75	
Level 2 (N 20g/m^2)	NH$_4$-N	−1.51	0.463	31.38		10.12
	NO$_3$-N	−1.23	15.62		17.38	
Level 3 (N 25g/m^2)	NH$_4$-N	+0.99	0.532	33.85		10.12
	NO$_3$-N	−0.6	16.77		18.01	

5 Summary

A Nitrogen-2D model to predict the nitrogen dynamics during treated wastewater application to FILTER systems was developed. The model simulates both water movement and solute transformations. The model numerically solves the Richards' equation for saturated-unsaturated water flow and the Fickian-based advection-dispersion equation for nitrogen transport, in which the transformations of nitrogen in both organic and inorganic matters are considered. The transport equation includes the linear equilibrium adsorption, zero-order production, and first-order degradation. The effects of temperature, water content, C/N ratio, and pH on the nitrogen transformation are considered in the model. The governing flow and transport equations are solved using a Galerkin type linear finite element scheme.

Model simulations were initially tested using field data from FILTER plots during 12 effluent application events of a summer cropping season and then used the model for the analysis of water flow and nitrogen transport in another effluent application season. The model accurately predicted the changes in the groundwater table depths in the FILTER plots with subsurface drains placed at 0.85 m, as well as the concentration of nitrate-nitrogen and ammonium-nitrogen in the subsurface drainage water outflows from the plots. The model, using the same water flow and nitrogen transformation parameters, was used for the simulation of water and nitrogen transport and transformation of a lysimeter experiment. The model accurately predicted the changes of water content and the concentration of nitrate-nitrogen and ammonium-nitrogen in the soil profile in 86 days. The model predicted water content and nitrogen distribution very well. It is indicate that the model can reasonably represent the water flow and nitrogen transport and transformation under the sewage irrigation conditions. We used the model to analyse nitrogen balance with difference irrigation strategy and the results can be used as a reference or guidance for sewage irrigation.

References

[1] Andren O., Paustian K.,1987. Barley straw decomposition in the filed – a comparison of models. Ecology, 68, 1190-1200.

[2] Carbon F., G. Girard, E. Ledoux.1991. Modeling of the nitrogen cycle in farm land areas. Fertilizer Research, 27. 161-169.

[3] Feddes R. A., Kowalik P. J., Zaradny H. 1978. Simulation of field water use and crop yield. Pudok, Wageningen.

[4] Genuchten MTh. Van. 1980. A closed-form equation for predicting the hydraulic conductivity of unsaturated soils. Soil Sci. Soc. Am. J, 44:892-898.

[5] J Simunek, T Vogel, M Th van Genuchten. The SWMS-2D for Simulating Water Flow and Solute Transport in Two-Dimensional Variably Saturated Media: Version 1.21. Research Report No. 132. US Salinity Laboratory Agricultural Research Service. US Department of Agriculture Riverside, California, 1994.

[6] J Simunek, M Sejna, M Th van Genuchten. April 1999. The HYDRUS-2D Software Package for Simulating the Two-Dimensional Movement of Water , Heat, and Multiple Solutes in Variably-Saturated Media: Version 2.0. US Salinity Laboratory Agricultural Research Service. US Department OF Agriculture Riverside, California.

[7] Jayawardane N. S.,1995. Wastewater treatment and reuse through irrigation, with special reference to the Murray Darling Basin and adjacent coastal areas. CSIRO, Div. Water Resources, Griffith NSW, Divisional Report 95.1.

[8] Johnsson H., Bergstrom L., Jansson P-E, Paustian. 1987. Simulated nitrogen dynamics and losses in a layered agricultural soil. Agriculture, Ecosystems and Environment, 18, 333-356.

[9] Kjoller A. Struwe S. 1982. Microfungi in ecosystems: fungal occurrence and activity in litter and soil. Oikos, 39: 389-422.

[10] L. Wu, M. B. McGechan. 1998. A review of carbon and nitrogen processes in four soil nitrogen dynamics models. Journey of Agriculture Engineering Research, 69; 279-305.

[11] Pescod MD. Wastewater treatment and use in agriculture. FAO irrigation and drainage paper 47. Soil Survey Staff. 1975. Soil Taxonomy: A basic system of soil classification for making and interpreting soil surveys. USDA Agricultural Handbook No. 436. Government Printer, Washington DC.

[12] Rijtema P. E. and J. G. Kroes. 1991. Some results of nitrogen simulation with the model ANIMO. Fertilizer Research, 27, 189-198.

[13] Stace HCT, Hubble GD, Brewer R, Northcote KH, Sleeman JR. Mulcahy MT, and Hallsworth EG. 1968. A handbook of Australian soils. Rellim Technical Publications, Glenside, S.A.

[14] Vinten A J A, Castle K, Arah J R M. 1996. Field evaluation of models of denitrification linked to nitrate leaching of aggregated soil. European Journal of Soil Science, 47, 305-317.

[15] Williams J. R., R. C. Izaurralde. 2005. The APEX Model. Texas A&M Blackland Research Center Temple, BRC Report.

[16] Zhang Y. F., W. Z. Zhang, Y. K. Shen, P. B. Lui, Sh. Y. Feng, and J.G. Shui. 1977. The transformation, transport, and loss of nitrogen in drainage farmland, Geology University Press, PP 184.

Seawater Intrusion and Land Subsidence Caused by Groundwater Overpumping in China

Wu Jichun Xue Yuqun Shi Xiaoqing
(Department of Hydrosciences, Nanjing University, Nanjing, China, 210093)

1 Seawater Intrusion Caused by Groundwater Overpumping

Seawater intrusion has long been a common topic at international conference (such as the Seawater Intrusion Meeting and the Seawater Intrusion in Coastal Area). A great number of papers have been published on this subject. In China, Seawater intrusion was observed in coastal aquifers as early as the 1960s. At present, it covers an area of more than 2,000 km^2 in China. One of the serious seawater intrusion occurred in Shandong Province, where the seawater intrusion area have increased to 1,773.6 km^2 in 2002, and it is continuing to extend in the coastal area of Laizhou Bay.

1.1 The Distribution of Seawater Intrusion

Seawater intrusion, caused by excessive exploitation of groundwater, has mainly occurred in the coastal areas of the following nine provinces in China: Liaoning, Hebei, Tianjin, Shandong, Jiangsu, Shanghai, Zhejiang, Hainan and Guangxi, of which Liaoning and Shandong are the most serious areas. Based on relevant data, the area of seawater intrusion of Shandong had totaled 1,773.6 km^2 by 2002 (seawater intrusion area had expanded to 601.31 km^2 in the coasts of Yantai Development Zone, Laizhou, Zhaoyuan, Longkou, Penglai, Zhifu, Haiyang, Laiyang, and expanded towards inland at the rate of 12% per year). In Liaoning, the total area of seawater intrusion reached 766 km^2 in the early 1990s. In 1994, seawater intrusion area had totaled 274.0 km^2 in Dalian. The seawater intrusion areas in other provinces are not large, ranging from several square kilometers to dozens of square kilometers. Taking Beihai City in Guangxi for example, seawater intrusion occurred mostly in certain coastal regions of Haicheng District. In March 1993, the seawater intrusion area was approximately 3 km^2. By the end of 1994, evidence of seawater intrusion was detected in coastal areas of Qiaogang Town in the southern part of the city. Take Hainan for another example. Seawater intrusion was discovered only in Xinying Bay, and the total area of seawater intrusion was 20 km^2 in 1999, covering the Xinying Town — Huangmu Village, Baimajing Town — Nan'an Village — Gongtangxia Village regions. According to general statistic, the total area of seawater intrusion caused by excessive groundwater exploitation exceeds 2,000 km^2 in China. From the perspective of region distribution, seawater intrusion concentrates on the following districts: ① areas surrounding Pohai Sea, including the coastal areas of the provinces of Shandong, Hebei, Liaoning and Tianjin, ② the coastal area of the Yangtze River Delta, such as Nantong, Shanghai, Ningbo and Wenzhou, and ③ the coastal area of southern China, such as Beihai and Haikou.

1.2 Characteristics of Seawater Intrusion

Based on an analysis of available data, the development of seawater intrusion in China has several main characteristics.

From spot intrusion to area intrusion. Seawater intrusion originated from an isolated spot intrusion, no more than 0.5 km^2, and then expanded gradually to form areas, and spread to the whole coast as area intrusion. Spot intrusions can hardly be seen in the Longkou—Laizhou region, but can be seen in Jimo, Haiyang, Mouping and so on. Seawater intrusion occurred in Longkou in 1977. The intrusion region was divided into 4 parts in 1984, two of which were no more than 6.4 km^2, but the intrusion spot near Longkou town and other two intrusion spots adjacent to it had connected together in 1988. Seawater intrusion spread to the western coast of Longkou, then developed to affect shores of the whole Longkou in 1994. The seawater intrusion occurred in Laizhou City in 1976, and the intrusion area was 71.1 km^2, thenceforward connected together and affected the whole coastal area of Laizhou City. The whole intrusion of Longkou — Zhaoyuan — Laizhou coastal areas had already formed into one piece.

Restricted by the geological conditions, there are several intrusion types.

(1)Area intrusion, occurs generally in Quaternary deposit areas or where fissures distribute homogeneously and with a uniform hydraulic connection densely.

(2)Finger intrusion, occurs generally in palaeo-channels. When high tide level occurs, seawater will run back for a long distance along recent riverbed (such as Baisha River in Mountain Laoshan, Qingdao, running back for 1,000~3,000 m), affecting groundwater quality on side banks.

(3)Veined intrusion, occurs generally in sparse structural fissure or fracture zones. Seawater intrusion is caused by fracture intrusion in gold mines on Sanshandao island, Laizhou City. Seawater intrusion in Nancheng district of Changdao and Longkou district of Qingdao are also caused by fracture zone intrusion, resulting in the pollution and abandonment of original water sources.

(4)Dendritic intrusion, occurs generally in karst aquifer systems. Seawater intrusion areas have relation with the dramatic decrease of groundwater level and the distribution of negative regions. The development of negative regions provides conditions for seawater intrusion, which develops along with negative regions but lags behind the development of negative regions. Once seawater intrusion occurs, its area hardly decreases along with the decrease of negative regions(see Table 1).

Table 1　Areas of Negative Regions and Seawater Intrusion in Longkou City, Shandong

(Data as of Jun)　　　　　　　　　　(Unit:km^2)

Year	Negative Regions	Seawater Intrusion Areas	Year	Negative Regions	Seawater Intrusion Areas
1984	109	64.5	1996	129	108.8
1988	146.5	78.4	1997	81.7	108.6
1989	177.5	85.7	1998	93.5	105
1990	178.7	88.7	1999	81.5	101
1991	137.9	89.4	2000	201	101
1992	227	91.5	2001	203	108
1993	218	96.4	2002	46	102
1994	224.5	103	2003	101	105
1995	225	100			

The specific distribution of seawater intrusion has relation with the strong pumping center. In the vicinity of coastal regions, seawater intrusion will occur in the areas of the strong pumping center and the depression cones. Generally, the interface between fresh and salt water originates on the side towards land adjacent to the strong pumping center, where seawater intrusion ends. If the strong pumping center moves to land, seawater intrusion will continue to move forward until a new balance forms. (Xue and Wu et al., 1993; Wu and Xue, 1993).

A wide transitional zone (mixing zone) exists between the sea water and the fresh water. The width of a transitional zone spans $1.5 \sim 3.5$ km^2, $2 \sim 6$ km^2, and over 7 km^2, respectively, in the cities of Longkou, Laizhou and Dalian. Until now, the very narrow transitional zone has not been detected between the seawater and the fresh water, which could be considered as a sharp interface.

The speed of seawater intrusion is affected by such factors as groundwater exploiting volume, exploiting mode and precipitation. The unrestrained exploitation of groundwater will result in the accelerated intrusion of seawater year by year. The typical zones are coastal areas of Shandong, and the cities of Dalian and Qinhuangdao. Take Laizhou as an example. The average annual growth of intrusion area were about 4 km^2 and 11 km^2 in the late 1970s and the early 1980s respectively, however it reached or exceeded about 30 km^2 in the mid-1980s, accordingly the seawater intrusion area developed from 15.8 km^2 in 1979 to 238.7 km^2 in 1989 (beyond 80% of total coastal area in the city). Take Longkou for another example. The average annual growth of intrusion area was about 4.6 km^2 in the mid-1980s, however it increased by 19.4 km^2 only in 1989, accordingly the seawater intrusion progressed from below 2 km^2 in 1979 to 85.7 km^2 in 1989. Dalian is also a good example. The seawater intrusion occurred in the mid-1960s and the area was only 4.2 km^2, then the intrusion progressed rapidly and the intrusion area was 178.5 km^2, 190.2 km^2 and 274.0 km^2 in 1981, 1986 and 1994, respectively. Nevertheless, the recognition timely of the seriousness of seawater intrusion, and taking effective measures, will retard seawater intrusion and even decrease its area. For example, some measures were taken by Laizhou in the early 1990s, which led to a considerable decrease in the spread of seawater intrusion. Seawater intrusion area was 260 km^2 in 2001, increased only 21.3 km^2 in the previous after 12 years. At present, seawater intrusion is under control. Several kinds of prevention and treatment measures were taken in Dalian in 1994, so the intrusion slowed down and the intrusion area decreased. According to statistic data from Dalian Environmental Monitor Station, the seawater intrusion area was only 161 km^2 in 2001. Therefore, seawater intrusion can be controlled as environment protection awareness increasing, parts of immediate and short-term economic benefits abandoned, and management of groundwater resources strengthened, water saved, drilling controlled, groundwater exploitation volume reduced, engineering measures taken to retain runoff into the sea and water storage capacity increased.

1.3 Prevention and Control of Seawater Intrusion

From the 1990s, Longkou City and Laizhou City and then the whole district of Pohai Sea, launched the research of prevention and control measures of seawater intrusion, and witnessed dramatic achievement. For example, seawater intrusion area in Longkou has been controlled at 100 km^2 in recent decades, without further development. Seawater intrusion has slowed down significantly in Laizhou and seawater intrusion of some places in Laizhou did not develop any longer in recent years. In Dalian Seawater intrusion tends to decrease in recent years.

The prevention and control of seawater intrusion encompasses the following aspects:

(1)Corresponding mechanisms of seawater intrusion by excessive exploitation of fresh water is explored through mechanism research of seawater intrusion.

(2)Mathematical models have been constructed in southeast coastal area of Laizhou Bay (such as Longkou, Laizhou and Yantai), Dalian in Liaoning Province and Beihai in Guangxi, respectively, which can forecast the seawater intrusion partially.

(3) Comprehensive measures are carried out in some areas of the Pohai Sea Region.

①Enhance the management of groundwater exploitation in the coastal regions. In order to exploit and utilize water resources rationally, strengthening the approval of water pumping permits, prohibiting driving deep wells in coastal areas, and controlling excessive groundwater exploitation in coastal areas are necessary.

②Implement the impoundment, source supply and recharging project. For the river flowing into the sea, with basin planning, overall arrangement and considering interests of upper, middle and lower courses, a medium and small sized impoundment project (including barrage brake and dam), assisting recharging and source supply project (including mini-type storage and seepage projects, such as ponds and canals, and recharging and source supply project, such as seepage ditches and walls) should be constructed, to utilize local flood during flood period for impounding and recharging groundwater furthest. In the Huangshui River basin comprehensive harnessing model (build reservoirs in upper course, impound and recharge layer upon layer in middle course, build underground reservoir in lower course) was taken in Longkou. On the basis of the underground reservoir, six central water supply source places were built in the Huangshui River drainage area to supply water for city zone and enterprises. At present, the daily water supply quantity is about 140×10^3 m^3, relieving the conflict between supply and demand of local water resources, controlling the development of seawater intrusion and sustaining local economic growth. Measures taken by Laizhou are similar to Longkou, building 10 dams in middle-lower course of each river in coastal regions in recent years, many infiltration wells, seepage ditches and ponds. These measures play a comprehensive role of cut-off, impounding, storage and drainage, with better effects, such as obvious groundwater table recovery and obvious slowness of intrusion speed (Liu Zhenfan and Meng Fanhai, 2003).

③The measures are as follows: construction of agricultural water conservation projects to introduce ecological agriculture, expansion farm water-saving irrigation actively, development of low voltage pipes' irrigation and mini-irrigation, plantation, improvement of the environment, development of dry farming and reducing groundwater extraction.

④To strengthen sea culture of land and management of saline enlargement, and to take measures to prohibit abundant seawater immerse inland and only to develop marine aquaculture within the specified shoals.

⑤Construction of inter basin and external-basin diversion projects, such as the "introducing Yellow River water to help Yantai" project and the eastern line of South-North Water Transfer Project.

2 Land Subsidence Caused by Groundwater Overpumping—An example of Su-Xi-Chang Area and Shanghai City

The excessive groundwater withdrawal in the Yangtze Delta has resulted in the serious geological disaster, for example the land subsidence. The land subsidence region in each city of the Yangtze Delta (south of the Yangtze River) has connected into a whole. The situation in Su-Xi-Chang area in Jiangsu Province and the Shanghai City is serious. In this paper, the Su-Xi-Chang area and the Shanghai City are treated as a whole studied area for the land subsidence studying. Based on the data from layered marks and observation wells, the relationship between the deformation and the groundwater level is analyzed.

2.1 General situation of the study area

In China, land subsidence mainly occurs in 17 provinces (cities) located in the eastern and middle regions, including Shanghai, Tianjin and Jiangsu, Hebei Provinces and so on. Yangtze Delta (South of the Yangtze River) included Shanghai city, Su-Xi-Chang area in Jiangsu province (including Suzhou, Wuxi and Changzhou Cities) and Hang-Jia-Hu area (including Hangzhou, Jiaxing and Huzhou Cities) in Zhejiang Province is one of the most serious subsidence area. The total area is about 26,830 km^2, in which the areas of Su-Xi-Chang, Hang-Jia-Hu area and Shanghai City are 14,000 km^2, 6,490 km^2 and 5,000 km^2, respectively. At present, the excessive exploitation of the groundwater leads to the groundwater depression cone getting across the provincial boundary, forming the huge regional depression cone. Accordingly the land subsidence has also connected into a whole, centered by the Shanghai, Su-Xi-Chang. Up to 2002, the maximum accumulative subsidence of Shanghai and Su-Xi-Chang are 2.63 m and 2.00 m respectively (Sun, 2002).

The selected studied area includes the Suzhou, Wuxi and Changzhou in Jiangsu Province and Shanghai City (except the Chongming, Changxing and Hengsha islands). The total area is about 17,000 km^2. It borders the Maodong plain and the east of the Mogan Mountain in the west, the East Ocean and Yellow Ocean in the east, the Yangtze River in the north, and is separated from the Zhejiang Province in the south.

The studied area is the accumulation plain of the Quaternary and the deposit thickness increases from west to east and from south to north. The average elevation is less than 6 m. The slope gradient is about 1/10,000. The Quaternary sediment is primarily composed of sand silt, medium-coarse sand and medium-coarse sand with gravels, with mild clay and clay interlayer. The interlayer of the clay and the sand constitutes 5 to 6 sedimentary rhythms in the vertical direction. In the studied area, there is mainly the pore water in the loose deposit. From the surface, it can be divided into the unconfined aquifer, the 1st, 2nd, 3rd, 4th and 5th confined aquifers. Note the division manner is different in Su-Xi-Chang area and Shanghai City. In Su-Xi-Chang area, the aquifer system is divided into the unconfined aquifer, the 1st, 2nd, 3rd and 4th confined aquifers. However, in Shanghai, it is separated into the unconfined aquifer, the 1st, 2nd, 3rd, 4th and 5th confined aquifers. According to the geological era, hydrodynamic condition and the formation type of each confined aquifer, the 2nd, 3rd, 4th and 5th confined aquifer in Shanghai correspond to the 1st, 2nd, 3rd and 4th confined aquifer in Su-Xi-Chang, respectively. For the unification, the division method in this paper is in accordance with that in Su-Xi-Chang area, that is, the 1st and the 2nd confined aquifers in Shanghai are combined and called the 1st confined aquifer, correspondingly the 3rd, 4th and 5th confined aquifers in Shanghai are named the 2nd, 3rd and 4th confined aquifers, respectively, is the hydrological profile of the studied area.

2.2 Groundwater exploitation and the land subsidence

The groundwater pumping is the main reason of the land subsidence in the studied area. Because of the different condition of the economic development and the aquifer system, the groundwater exploitation history, the main pumping layer and the pumping yield are different in different district. In 2000, the pumpage in Su-Xi-Chang area contributed 62% of total pumpage in Yangtze Delta (South of the Yangtze River). The ratio of the pumpage in Shanghai and Hang-Jia-Hu area is 17% and 21%, respectively (Zhang and Wei, 2005).

The large scale groundwater exploitation in Shanghai started in the 1950s. The 80.5% of the yield was from the 1st and 2nd confined aquifer, which resulted in the serious compress of the shallow soil layer. The larger area of the subsidence depression cone occurred in the concentrated pumping area. After 1966, the pumping layer was adjusted and the yield was decreased in order to control the subsidence in the downtown. The pumping yields from the 1st and 2nd confined aquifer were restricted and the yields from the 3rd and 4th confined aquifer were increased. As a result, the subsidence rate slowed down obviously, even having the rebound. In the late of the 1980s, the need for the groundwater increased due to the economic construction. In 2000, the pumping yields from the 3rd and 4th confined aquifer accounted for the 70% and 15% of the total yields in Shanghai. The subsidence rate in Shanghai was increased (Wei, 2002; Wei et al., 2005).

With the development of the economy and the increase of the population, groundwater extraction has been rapidly increased since the 1980s. The 1st and 2nd confined aquifers are the main aquifers under exploitation. The groundwater depression cone, with 30 km wide, 125 km long and larger than 5,000 km^2 area, was formed along the Hu-Ning railroad. Correspondingly, the regional land subsidence cone was formed gradually. Though the groundwater yields kept decreasing after the manage enforcement of the groundwater, the total yields was still larger. The pumping yield in Suzhou, Wuxi and Changzhou was 3×10^9 m^3 and the area of the subsidence depression cone with the accumulative subsidence larger than 0.2 m reached 5,000 km^2 in 2000 (NCCGS et al., 2003).

Based on the analysis on the development of land subsidence in the studied area, the basic features are listed in the following:

(1) Land subsidence took place more than 1/3 of the whole region. It often takes place in the area with confined aquifers except the bedrock.

(2) The development of land subsidence was coincident with the groundwater depression cone in the time and space. Taking as Su-Xi-Chang area for an example, the groundwater pumping focused on the three center cities: Suzhou, Wuxi and Changzhou. Thus the subsidence took place in the centers. After 1990s, with the increasing of groundwater exploration and groundwater level declining rapidly, the subsidence spread to the whole area and join together soon. The configuration of the subsidence cone in some places, such as the downtown of Suzhou and Changzhou and the west of Wuxi, is basically identical with the shape of the groundwater depression cone, which clearly indicated that there were strong relationship between the development of land subsidence and the groundwater level of the main pumping aquifer (Yu et al., 2004, 2006).

(3) The subsidence in studied area is the result of the consolidation of the sandy and soft soil strata due to the long-time excessive groundwater pumping of the confined aquifers.

Taking the extensometer of Qingliang Primary School in Changzhou for example, the clay layers above the 2nd confined aquifer, especially the clay layer whose water level depth from the surface is 35.40~92.80 m, are the main consolidation layer. The ratio of compression deformation for these clay layers is 57.15%. Notable, from 1984 to 2001, the compress subsidence proportion of the 2nd confined aquifer with the depth from 92.80 m to 107.80 m reaches 12.45% and the subsidence of the unit thickness is 5.29 mm, which demonstrate that the compress deformation of this sandy aquifer can not be ignored. In addition, the accumulative compress subsidence proportion of the 3rd confined aquifer, whose water level depth from the surface is from 117.65 m to 143.6 m, is 21.34%. Its accumulative compaction deformation from 1995 to 2003 is 44.03 mm, which is the maximum compaction deformation in the aquifer system. This clearly indicates that the compressibility of the sandy layers should not be ignored (Shi et al., 2006).

Table 2 is the deformation statistic of different layers in Shanghai during the period of 1980~2000. As shown in Table 1, the compression deformation of the 3rd confined aquifer had greatly effect on land subsidence. It is should be noted that the ratio of the deformation of the 3rd and 4th confined aquifer increased from 37.7% to 52.23%, which demonstrated that the deformation of the deep confined aquifers is most important on land subsidence in Shanghai.

Table 2 The deformation features of different layers in Shanghai downtown
(modified from Wei et al., 2005)

Soil layer	Average thickness (m)	Time period of 1980~1995				Time period of 1996~2000			
		Annual deformation rate (mm/a)	Accumulated deformation (mm)	Ratio (%)	Rank	Annual deformation rate (mm/a)	Accumulated deformation (mm)	Ratio (%)	Rank
1st aquitard	31	−1.96	−29.42	29.9	2	−4.95	−24.78	21.8	2
1st confined quifer	87	−1.82	−27.34	27.7	3	−3.09	−15.47	13.6	3
2nd confined quifer	40	−0.32	−4.82	4.9	5	−2.83	−14.14	12.4	4
3rd confined quifer	87	−3.13	−31.90	32.3	1	−11.21	−56.06	49.3	1
4th confined quifer	78	−0.36	−5.33	5.4	4	−0.67	−3.33	2.93	5

References

[1] Jichun Wu, Yuqun Xue, Peiming Liu, Jianji Wang, Qingbo Jiang and Hongwen Shi, Seawater Intrusion in the Coastal Areas of Laizhou Bay, China -2. Sea Water Intrusion Monitoring, Ground Water, 1993, Vol. 31, No. 5.pp 740-745.

[2] Liu Zhenfan and Meng Fanhai, The Coastal Underground Reservoir System Engineering in Longkou City, Journal of Kanchakexuejishu, 2003, No. 6,pp 47-52.(in Chinese)

[3] NCCGS(Nanjing Center, China Geological Survey), Geological Survey of Jiangsu Province, SIGS(Geological Survey of Shanghai), Geological Survey of Zhejiang Province, Investigation and estimation on groundwater resources and geological disasters on Yangtze Delta, pp265, 2003.(in Chinese)

[4] Shi X. Q, Xue Y. Q, Wu J. C., et al. Study on soil deformation properties of groundwater system in Changzhou area. Hydrogeology and Engineering Geology, 33(3): 1-6, 2006. (in Chinese)

[5] Sun W.S.. Investigation report on prevention and control of land subsidence in Yangtze Delta. In: Wei Z. X., Li Q. F. (Eds). Proceedings of the national symposium on land subsidence, Shanghai Institute of Geology Survey, shanghai, pp.1-12, 2002.(in Chinese)

[6] Wei Z. X., Yang G. F.and Yu J. Y., Stress-strain characteristics of the confined aquifer system and land subsidence controlling countermeasures in Shanghai. The Chinese Journal of Geological Hazard and Control, 16(1): 5-8. 2005. (in Chinese)

[7] Wei Z. X., Stress and strain analysis of the fourth artesian aquifer in Shanghai[J]. The Hydrology and Engineering Geology, Vol. 29, No. 1,1-4, 2002. (in Chinese)

[8] Yu J, Wu J. Q., Wang X. M., Yu Q. Research on the correlative prediction model with a regional decomposition base of the land subsidence in the Suzhou-Wuxi-Changzhou area[J]. Hydrogeology and Engineering Geology. 4:92-95. (in Chinese)

[9] Yu J., Wang X. M., Wu J. Q. Xie J. B., Characteristics of land subsidence and its remedial proposal in Suzhou-Wuxi- Changzhou area. Geological Journal of China Universities, 12(2): 179-184. 2006.(in Chinese)

[10] Yuqun Xue, Jichun Wu, Peiming Liu, Jianji Wang, Qingbo Jiang and Hongwen Shi, Sea Water Intrusion in the Coastal Areas of Laizhou Bay. China -1. Distribution of Sea Water Intrusion and Its Hydrochemical Characteristics, 1993, Ground Water, Vol. 31, No.4,pp 532-537.

[11] Zhang A. G. and Wei Z. X.eds, Land subsidence in China. Shanghai: Shanghai Scientific & Technical Publishrs, 2005.(in Chinese)

污染场地健康风险评价的理论和方法

陈鸿汉 谌宏伟 何江涛 刘菲 沈照理 韩冰 孙静

(中国地质大学水资源与环境学院，北京 100083)

污染场地健康风险评价指对已经或可能造成污染的工厂、加油站、地下储油罐、垃圾填埋场、废物堆放场等场地由于污染物质排放或泄漏对人体健康的危害程度进行概率估计，它是一项多学科交叉的复杂的系统工程。

20 世纪 80 年代以来，欧美国家在环境风险评价的理论基础上先后建立起了污染场地健康风险评价体系。美国环保局于 1980~1988 年先后颁布了《环境响应、补偿与义务综合法案》等作为响应污染物排放和突发污染事件的法律性文件[1]，并制定了一系列诸如《健康风险评价手册》、《场地治理调查和可行性分析指南》等风险评价导则[2,3]，形成了包括法律法规、导则指南和技术文件在内的一整套完善的污染场地健康风险评价体系。欧盟 16 国于 1994 年成立欧盟污染场地公共论坛，并于 1996 年完成污染场地风险评价协商行动指南[4]。加拿大、澳大利亚和芬兰等国基本沿用美国的风险评价方法[5,6]，同时构建了适合本国实际的健康风险评价体系。

我国在 20 世纪 90 年代，开始了以介绍和应用国外研究成果为主的环境风险评价研究[7,8]，但大部分集中在事前风险评价。同时，我国环境保护法和环境影响评价法也只对规划和建设项目开展环境影响评价作出了规定，尚未涉及污染场地健康风险评价方面的内容。

本文综合了目前国外污染场地健康风险评价的研究成果，探讨污染场地健康风险评价方法，并对构建适合我国实际的污染场地健康风险评价体系开展探讨。

1 健康风险评价基础理论

1.1 人体污染物摄取方式和机制

人体摄取污染物质的途径主要包括：口、呼吸和皮肤接触。通常采用不同类型剂量来表示污染物质进入人体各个阶段的数量[6]。无论通过何种途径，污染物质只有最终进入到人体血液中才会对人体健康产生影响，因此以污染物质透过肺泡呼吸膜、胃肠壁黏膜和皮肤进入血液的数量为依据。污染物的运动以扩散作用为主，如果将呼吸膜等以均质层对待，并假设介质和其中的污染物质互为独立扩散过程，且污染物质对这些扩散层不产生损伤，可以用费克第一定律(扩散定律)估算污染物质进入人体血液的数量。扩散通量可表示为：

$$J = K_m \Delta C \tag{1}$$

式中：J 为单位时间单位膜或皮肤表面积上通过的污染物质数量；ΔC 为膜或皮肤等两侧污染物质浓度差，即浓度梯度；K_m 为膜或皮肤的渗透系数，可表述为膜或皮肤 2 介质分配系数($K_{m/v}$)、污染物质在膜或皮肤中的扩散系数(D) 和扩散距离(l) 的函数：

$$K_m = K_{m/v} D l \tag{2}$$

1.2 剂量—反应关系

污染物质对人体产生的不良效应以剂量—反应关系表示。对于非致癌物质通常认为存在阈值现象。对于致癌和致突变物质，一般认为无阈值现象，即任意剂量的暴露均可能产生负面健康效应。

1.2.1 非致癌效应

非致癌效应的阈值的表征方法有：不可见有害作用水平(NOAEL)、最低可见有害作用水平

(*LOAEL*)和基准剂量(*BMD*)。传统上主要以实验所得的 *NOAEL* 和 *LOAEL* 表示，但这两种表述方法没有考虑剂量—反应曲线的特征和斜率，不能真实地表达受试物的毒性与效应，有逐渐被 *BMD* 取代的趋势。

非致癌风险的标准建议值根据参考剂量/浓度(*RfD/RfC*)、可容忍日摄取量(*TDI*)和可接受日摄取量(*ADI*)等而定。美国环保局考虑人群个体差异、动物实验数据应用到人体、短期实验数据用于长期暴露以及由 *LOAEL* 代替 *NOAEL* 所带来的不确定性，采用式(3)计算参考剂量：

$$RfD = NOAEL/(UF1 \times UF2 \times UF3 \times UF4 \times MF) \tag{3}$$

式中：*UF* 为不确定因子，取值 1~10；*MF* 为修正因子，取值为 1~10。

1.2.2 致癌效应

致癌效应的剂量—反应关系是以各种关于剂量和反应的定量研究为基础建立的。由于人体在实际环境中的暴露水平通常较低，而实验学或流行病学研究中的剂量相对较高，因此在估计人体实际暴露情形下的剂量—反应关系时，常常利用实验获取的剂量—反应关系数据推测低剂量条件下的剂量—反应关系，称为低剂量外推法。

低剂量外推法包括线性和非线性两种模型。模型的选择主要基于污染物的作用模式。当作用模式信息显示低于出发点剂量的剂量—反应曲线可能为线性时，则选择线性模型。如污染物为 DNA 作用物或具有直接的诱导突变作用，其剂量—反应曲线常常为线性。当污染物的作用模式不确定时，线性模型为默认模型。当充分的证据表明污染物的作用模式为非线性，且该物质不具有诱导突变作用时，可采用非线性模型。由于某些物质同时对不同的器官具有致癌作用，则可根据作用模式的不同，分别采用线性和非线性模型。此外，当有证据证实在不同的剂量区间内，污染物对同一器官的作用模式分别为线性和非线性时，可以结合使用线性和非线性模型。

2 健康风险评价方法

2.1 数据收集和分析

数据收集包括：①场地背景资料：主要包括场地物理特征、利用历史、布局等，它是暴露评估中暴露背景以及建立污染物迁移转化模型的资料来源；②场地污染状况：主要指场地污染历史和现状；③与污染物有关的资料：污染类型、污染物种类、污染物物理化学性质和毒理学证据等；④与暴露人群有关的资料：人群分布、结构、生活方式等。

2.2 暴露评估

暴露评估指定量或定性估计暴露量、暴露频率、暴露期和暴露方式。

呼吸途径和饮食途径一般采用潜在剂量进行估算，皮肤接触途径采用吸收剂量估算。

2.2.1 呼吸途径

呼吸挥发性气体：

$$Intake = \frac{C_a \times IR \times ET \times EF \times ED}{BW \times AT} \tag{4}$$

呼吸可吸入颗粒物：

$$Intake = \frac{C_p \times FP \times IR \times ET \times EF \times ED}{BW \times AT} \tag{5}$$

式(4)、式(5)中：*Intake* 为单位时间单位体重污染物摄取量，mg/(kg·d)；C_a 为空气中挥发性气体的浓度，mg/m³；*IR* 为摄取速率，m³/h；*ET* 为暴露时间，h/d；*EF* 为暴露频率，d/a；*ED* 为暴露期，a；*BW* 为人群平均质量，kg；*AT* 为平均暴露时间，d；C_p 为空气中可吸入颗粒物含量，kg/m³；*FP* 为可吸入颗粒物中污染物含量，mg/kg，对于致癌物质，为人群平均寿命，对于非致癌物质，为暴露期。

2.2.2 饮食途径

饮水：

$$Intake = \frac{C_W \times IR \times EF \times ED}{BW \times AT} \tag{6}$$

食物：

$$Intake = \frac{C_F \times IR \times FI \times EF \times ED}{BW \times AT} \tag{7}$$

式(6)、式(7)中：C_W 为水中污染物浓度，mg/L；C_F 为食物中污染物含量，mg/kg；FI 为污染食物占总食物的比例；IR 为摄取速率(水：L/d；食物：kg/meal)；EF 为暴露频率(水：d/a；食物：meal/a)；其他符号含义同前。

2.2.3 皮肤接触途径

皮肤接触污染水体：

$$Absorbed\ Dose = \frac{K_p^w \times C_W \times SA \times ET \times EF \times ED \times CF}{BW \times AT} \tag{8}$$

皮肤接触污染土壤：

$$Absorbed\ Dose = \frac{C_S \times F_{adh} \times SA \times ABS \times EF \times ED \times CF}{BW \times AT} \tag{9}$$

皮肤接触污染空气：

$$Absorbed\ Dose = \frac{K_p^a \times C_a \times SA \times ET \times EF \times ED}{BW \times AT} \tag{10}$$

式(8)~式(10)中：$Absorbed\ Dose$ 为单位时间单位体重皮肤吸收污染物数量，mg/(kg·d)；K_p^w、K_p^a 分别为与水和空气接触时污染物在皮肤中的渗透系数，cm/h；C_S 为土壤中污染物浓度，mg/kg；ABS 为皮肤对污染物的吸收分数；SA 为与污染水体、土壤和空气接触的皮肤表面积，cm²；F_{adh} 为土壤对皮肤的黏附系数；CF 为单位转换因子(水：1 L/(1 000 cm³)；土壤：10^{-6} kg/mg)；EF 为暴露频率(水：d/a；土壤：events/a)；其他符号含义同前。

2.3 毒性评估

毒性评估是指利用场地目标污染物对暴露人群产生负面效应的可能证据，估计人群对污染物的暴露程度和产生负面效应的可能性之间的关系[2]。一般分为危害识别和剂量—反应评估两个步骤。

本文主要研究长期暴露于小剂量化学污染物引起的致癌和非致癌风险。基于短期暴露于较大剂量污染物和长期暴露于小剂量污染物所带来的致癌作用具有等价效应的假设[6]，致癌风险评估常采用人体终生暴露可能造成的健康风险表示。美国环保局将污染物质的致癌毒性分为 A、B、C、D、E 5 大类，并用剂量—反应曲线所确定的斜率因子这一标准建议值表示人体终生暴露于一定剂量某种污染物质而产生致癌效应的最大概率[2]。

2.4 风险表征

2.4.1 风险估算

以致癌风险和非致癌危害指数表示。目前国外通常采用单污染物风险和多污染物总风险以及多暴露途径综合健康风险 3 种方式表示。

致癌风险：

当 $Risk<0.01$ 时 $Risk=CDI \times SF$

当 $Risk>0.01$ 时 $Risk=1-\exp(-CDI \times SF)$ \qquad (11)

非致癌危害指数：

$$HQ = Intake\ 或\ Absorbed\ Dose/RfD \tag{12}$$

多污染物总风险：为某一暴露途径各污染物风险之和。

致癌总风险:

$$(Risk)_T = \sum (Risk)_i \tag{13}$$

非致癌总危害指数:

$$HI = \sum HQ_i \tag{14}$$

式(11)~式(14)中: $Risk$ 为致癌风险, 表示人群癌症发生的概率, 通常以一定数量人口出现癌症患者的个体数表示; CDI 为人体终生暴露于致癌物质的单位时间单位体重的平均日摄取量, mg/(kg·d); SF 为斜率因子, kg/(mg·d); HQ 和 HI 分别为单污染物和多污染物的非致癌危害指数, 其数值的大小表示风险的大小, 当小于1时, 认为风险较小或可以忽略, 当大于1时, 认为存在风险。

综合健康风险: 为各暴露途径总风险之和。

2.4.2 不确定性分析

不确定性来源于风险评价的各个阶段, 野外取样、实验分析、模型参数获取、模型的适用性和假设、毒理学数据等均存在客观和主观的不确定因素。

2.4.3 风险概述

风险概述指客观地表述场地风险, 充分分析风险评价的不确定性程度, 承认风险的相对性, 科学地指导场地污染防治决策。

3 评价方法的探讨

3.1 叠加风险

计算由目标污染场地造成的实际人群健康风险, 是叠加在其他污染场地或背景污染产生的人群健康风险之上的风险。具体计算如下:

(1)以实际监测资料评价现实健康风险, 直接用环境介质的监测浓度减去背景浓度后所得浓度值计算人体摄取量。空气介质, 可用污染场地的监测浓度减去其主风向上游的监测浓度; 土壤介质, 用污染土壤的浓度值减去四周土壤监测浓度平均值; 地下水, 用场地地下水监测浓度减去其上游监测浓度。

(2)采用模型预测浓度计算未来健康风险, 环境介质的初始污染物浓度应用初始监测浓度减去背景或上游污染物浓度。

在经过上述数据处理后, 计算所得风险为目标污染场地的实际风险, 即叠加风险。

3.2 多暴露途径同种污染物累计健康风险

目前国外污染场地健康风险评价方法主要以单污染物风险、多污染物总风险和多暴露途径综合风险。笔者在评价实例研究中发现, 对于多暴露途径情形, 应增加计算各暴露途径同种污染物对同一人群的累计健康风险。这不仅可以排除不同污染物之间拮抗和协同作用对评价结果的影响, 还可以清晰地表达某种污染物对人群的累计风险, 有利于有针对地制定污染治理标准。具体计算如下:

多暴露途径同种污染物的累计致癌风险:

$$Risk_T^A = \sum_{i=1}^{n} Risk_i^A \tag{15}$$

多暴露途径同种污染物的累计非致癌危害指数:

$$HI_T^A = \sum_{i=1}^{n} HQ_i^A \tag{16}$$

式(15)、式(16)中: $Risk_i^A$ 为暴露途径 i 中 A 污染物的致癌风险; HQ_i^A 为途径 i 中 A 污染物的非致癌危害指数。

4 中国开展污染场地健康风险评价的相关问题

(1)法律不健全,污染调查工作开展困难。迄今为止,我国尚无"污染场地健康风险评价"的法律法规。

(2)场地污染监测体系不完善,缺乏对场地污染历史和现状的全面了解。

(3)对场地目标污染物质的监测不全面,特别是有机污染物。同时我国环境质量标准中所列的污染物也不全面。这种状况将导致不能充分、全面评价污染场地的人群健康风险。

(4)对污染物质的环境行为认识不够。

(5)污染物质对人体的致病机理的研究还显得相对薄弱。

5 中国构建污染场地健康风险评价体系的建议

(1)尽快制定污染场地健康风险评价的法律法规,从国家法律层面上定污染场地健康风险评价的地位,使之有法可依。

(2)选择典型污染场地开展污染场地健康风险评价试点,探索风险评价方法,为污染场地健康风险评价提供实践经验。

(3)借鉴国外风险评价的经验,逐步建立和完善污染场地健康风险评价指南和技术细则。

(4)完善监测体系,着手建立土壤、地表水、空气污染状况与污染物物理化学性质和毒性数据库,为风险评价提供充分的数据基础。

(5)加强环境科学和医学等学科之间的交流,开展污染物环境行为和污染物致毒机理等方面的研究,为健康风险评价提供理论依据。

(6)污染场地风险评价体系的建立应结合健康风险和生态安全开展相关技术与方法研究。

参 考 文 献

[1] Congress of United States. Comprehensive environmental response, compensation and liability act [EB/ OL] . http://www. epa. gov , 1980.

[2] U.S.EPA. Risk assessment guidance for superfund : Human health evaluation manual [EB/ OL] . http :// www. epa. gov ,1989.

[3] U.S.EPA. Guidance for conducting remedial investigation and feasibility studies under CERCLA[EB/ OL] . Office of Emergency and Remedial Response, Washington DC. http ://www. epa. gov , 1988.

[4] COLIN C F. Assessing risk from contaminated sites: Policy and practice in 16 European countries [J] . Land Contamination and Reclamation, 1999 , 7 (2) : 33254.

[5] National Environmental Protection Council (NEPC). Guideline on health risk assessment met hydrology [EB/ OL]. http:/ /www. ephc. gov. au , 1999.

[6] Canadian Council of Ministers of the Environment (CCME) .Canada2 wide standards for pet roleum hydrocarbons in soil [EB/ OL] . http :/ / www. ccme. ca , 2001 : 128.

[7] Zhong Zhenglin, Zeng Guangming, Yang Chunping. A summary of environmental risk assessment research[J] . Environment and Development, 1998, 13 (1) : 39241.

[8] Hu Erbang. Technology and met hods of environmental risk assessment [M]. Beijing: China Environmental Science Press , 2000 : 12482.

华北平原地下水超采现状及对策

马凤山

(中国科学院地质与地球物理研究所，北京 100029)

黄河以北的华北平原面积 $13.6×10^4 km^2$，包括了北京、天津、河北、山东、河南五省市，人口约 1.11 亿人。该地区属季节性干旱半湿润大陆性季风气候区。多年平均降水量 500～600 mm。蒸发量大于降雨量 3.9 倍。春旱夏涝，秋冬又旱，旱涝交替。历史上平原排水出路集中在海河末梢天津，平原有许多洼地，自然排水不畅。

华北平原东部形成大面积地下咸水区，总面积 8.39 万 km^2，占平原面积的 61.7%。多年平均水资源总量约为 371.85 亿 m^3，人均水资源占有量仅为 335 m^3，不足全国平均水平的 1/6、世界的 1/24。该区域内地表水时空分布不均，资源短缺，海河流域地表水资源开发利用程度已经超过了 90%。

1 区域水文地质特征

按照华北平原地下水资源分布的特点，可将黄河以北华北平原区划分为山前洪积、冲洪积平原，中部冲积平原和滨海平原三种类型。华北平原第四系是一套砂泥多层交叠的复合地层，含水层岩性、结构、厚度等具有水平变化规律。在山前平原含水层呈扇状结构，扇轴含水层岩性以砾石、卵石为主，厚度大；扇间含水层粒度变细，厚度变薄。在中部平原含水层逐渐过渡为湖相沉积穿插河流沉积的舌状结构，含水层岩性以中细砂为主，厚度在靠山前平原方向变薄，向滨海方向又略变厚。向东部、南部的滨海平原含水层又过渡为湖积的岛状结构，含水层岩性以粉细砂为主，厚度又变薄。华北平原第四系是一套几何形态复杂的多种沉积类型交叉叠置的含水岩系。

依据地层结构特点，在平面上又划分了单层结构区和多层结构区。多层结构区将第四系含水岩系自上而下划分为 4 个含水层组。

山前洪积、冲洪积平原城市大都坐落于太行山和燕山山麓地带，是一个全淡水区，包括邢台市、邯郸市、石家庄市、保定市、北京市、唐山市等一系列特大、大及中小城市。该区具有补给多元化、垂向水力联系好、排水畅通、水质良好、地下水资源丰富的特点。对地下水的开采以第一、二含水组为主，地下水采补平衡关系打破得较晚。

中部冲积平原城市坐落于海河流域中部的冲积、冲湖积平原，是一个有咸水区。包括廊坊市、衡水市、沧州市等中小城市。该区水力坡度减小，渗流微弱，埋深变浅，蒸发变强，以垂向水循环交替为主，咸、淡水并存，地下水资源一般。主要开采第一、三含水组和深层的地下淡水。地下水采补平衡关系的打破与地区地下水灌溉农业和城市的快速发展有关。

滨海平原城市坐落于滨海冲积、海积平原，包括天津市、秦皇岛市等。地势低平，水力坡度仅为 0.10‰～0.25‰，导水系数一般小于 50 m^2/d，海相地层较发育，浅层潜水-微承压水基本为咸水，仅局部地段有薄层淡水透镜体，为咸水分布区，但深层为一淡水地下水系统。具有排水不畅、地下水埋深小、蒸发强烈、矿化度大、水质差、地下水资源缺乏的特点。以开采深层地下淡水为主，同时利用有限的浅层地下淡水和微咸水。因此，开采地下水引发的环境问题出现得比较早。

2 地下水资源与开发利用

最新的调查评价结果表明，华北平原地下水天然资源量为 $227×10^8 m^3/a$，其中，浅层地下水

开采资源量 $168\times10^8\,m^3/a$，深层地下水可采资源量 $24\times10^8\,m^3/a$。

2000 年华北平原地下水开采量为 $212\times10^8\,m^3$，其中浅层地下水开采量为 $178.4\times10^8\,m^3$，占地下水总开采量的 84.2%，多集中在全淡水区；深层地下水开采量为 $33.6\times10^8\,m^3$，占地下水总开采量的 15.8%，主要集中在有咸水区。浅层地下水开采程度为 106%，仅超采 $10.11\times10^8\,m^3/a$，基本处于采补均衡，但开发利用在平面和垂向上分布极不均匀。河北、北京等山前地区近 30 年来浅层地下水位普遍下降 20～40 m，超采严重；东部平原、沿黄河地区浅层地下水开采潜力较大。

全区深层地下水开发利用程度已达 139%，除河南、山东部分地区开采程度较低外，其余多数地区严重超采，致使华北平原大部分深层地下水头低于海平面，以地下水封闭的 0 m 等值线圈定的低于海平面范围为 $76\,732\,km^2$，占平原总面积的 55%。

地下水作为主要供水水源，河北省 2000 年地下水开采程度达到 132.31%，超采约 $31.41\times10^8\,m^3/a$，1975 年以来累计消耗地下水储存量超过 $500\times10^8\,m^3/a$；北京市 2000 年地下水开采程度 105.16%，超采 $1.22\times10^8\,m^3/a$，1961 年以来累计消耗地下水储存量 $60\times10^8\,m^3/a$ 以上(据北京市地质调查研究院 2003 年完成的"首都地区地下水资源和环境调查评价"推算)；天津市地下水资源贫乏，2000 年开采程度为 73.51%，但深层地下水开采程度达到了 159.96%，深层地下水超采近 $11.09\times10^8\,m^3/a$。山东、河南两省地下水开发利用程度也都在 75% 以上。由于长期超采，华北平原地下水位呈持续下降。

华北平原城市化地区地下水超采严重，范围已近 $9\times10^4\,km^2$。浅层地下水降落漏斗分布于山前沿线中心城市，已形成了以北京、石家庄、保定、邢台、邯郸、唐山为中心，总面积达 $4.1\times10^4\,km^2$ 的浅层地下水漏斗区，其中 $1\times10^4\,km^2$ 范围的含水层已疏干。同时，也形成了以天津、衡水、沧州、廊坊等多个城市为中心、面积达 $5.6\times10^4\,km^2$ 连成一体的深层地下水漏斗区。

由于华北平原大规模开采深层地下水，在一定程度上激发了深层地下水越流补给和侧向径流的能力，如天津地区侧向径流补给和相邻含水层的越流补给可达 50%～60%。而且区域深层地下水头持续下降，导致华北平原地区地下水降落漏斗面积逐年扩大、加深，形成了多个深层地下水降落漏斗复合体。

长期持续的地下水过量开采，华北平原在天津、沧州和北京东北郊分别形成三个沉降中心。最新调查表明(中国 2005 年地质环境公报)，华北平原不同区域的沉降中心仍在不断发展，且有连成一片的趋势。北京地区主要沉降中心为东八里庄—大郊亭、昌平沙河—八仙庄、大兴等地区，最大累计沉降量为 798 mm；天津地区主要沉降中心为塘沽、汉沽、市区等地区，最大累计沉降量为 3 187 mm；河北地区主要沉降中心为沧州、任丘等地区，最大累计沉降量为 2 457 mm；山东德州沉降区最大累计沉降量达 936 mm。沿海一带出现负标高地区 20 km^2，风暴潮灾害非常严重。沧州地面沉降区伴生出现了 20 多条地裂缝，最长达 4 km。

3 地下水资源可持续开发利用对策

3.1 调整地下水开采布局，实施含水层恢复与保护工程

由中国地质调查局组织实施的"华北平原地下水可持续利用调查评价"项目调查成果显示，华北平原地下水漏斗中心区(深层地下水头埋深大于 50 m)已不具备开采潜力，需严格控制开采。华北平原应积极开发利用具有较大潜力的浅层咸水、微咸水资源以及浅部薄层淡水资源，适当调整或压缩开采深层地下水资源，以控制降落漏斗的扩大。同时利用山前冲洪积扇倾斜平原地下水的良好蓄水区和干涸河道地下水调蓄的有利入渗场所，在南水北调工程实施后，压缩开采地下水，实施地下水自然调蓄，涵养地下水。该项目通过综合研究得出华北平原深层地下水环境约束条件的临界界限值为水位埋深 50 m，并将 70 m 作为严格控制界限，以防止地面沉降等环境问题的加剧。

3.2 实施南水北调，增加新的供水水源

根据南水北调规划设计确定的调水规模，中线工程从丹江口水库引水，年平均引水 $130\times10^8\,m^3$，

过黄河流量为 $70 \times 10^8 \sim 75 \times 10^8 \mathrm{m}^3/\mathrm{a}$，可部分缓解京广铁路沿线城市缺水；东线工程主要供水目标为黄淮海平原东部和山东半岛，一期过黄河流量为 $50 \mathrm{m}^3/\mathrm{s}$，二期过黄河流量为 $100 \mathrm{m}^3/\mathrm{s}$，三期过黄河流量为 $200 \mathrm{m}^3/\mathrm{s}$，可作为天津市的补充水源。

3.3 实施地下水人工调蓄，建立地下水人工调蓄示范工程

南水北调工程实施后，应充分利用当地地下水超采腾空的地下库容，对当地水和外调水进行调蓄。严格控制和减少地下水的开采量，不允许地下水继续超采。通过涵养和回灌逐步恢复地下水的战略储量。

3.4 加强对水资源的综合管理，提高水资源的利用效率

节约用水、防治水污染是目前解决华北地区水资源紧张的最有效途径，而造成华北平原地区水资源浪费和污染的最主要原因是管理不善，人们缺乏保护水资源的意识。

政府及相关职能部门要加强对水资源管理的执法力度和宣传力度，提高人们保护水资源的意识，采取措施，提高水资源的利用效率；如果能将城市供水损失率降低 5%，农业用水中渗水、漏水、不合理的灌溉而损失的水减少 15%，整个华北平原地区每年相当于增加水资源近 $180 \times 10^8 \mathrm{m}^3$。

3.5 大力发展节水农业

大力发展节水农业，积极推广耐旱作物，改革灌溉技术，如实行管道输水、喷灌、滴灌，比大水漫灌节约用水 $1/2 \sim 2/3$。

3.6 调整地下水开采布局，积极开发利用浅层淡水和微咸水资源

开发利用当地微咸水。南水北调通水以后，微咸水区的农业供水条件改善不大，仍有必要利用微咸水。因此，咸水区要相机调引部分淡水，采取咸淡轮灌、污咸混灌方式，带动微咸水的开发利用。

3.7 加强城市地下水水源地储备

通过"华北平原地下水可持续利用调查评价"项目调查，依据地下水开采潜力大小将华北平原地区依次划分为 5 个区：可扩大开采区、可适度扩大开采区、可维持现状开采区、适度控制开采区和严禁开采区。其中，浅层地下水资源最匮乏的严禁开采区占该地区总面积的 12.3%，包括石家庄、保定等城市。北京位于适度控制开采区。

在一定的地质环境约束条件下，为支撑社会经济发展，需要在一定时期临时动用应急地下水存量。经计算分析，华北平原区 2010 年南水北调实施前可提供地下水应急开采量为 $75.01 \times 10^8 \mathrm{m}^3/\mathrm{a}$，可在一定程度上缓解该地区水资源供求压力。

为保障京津冀主要城市的供水安全，在调查评价的基础上，中国地质调查局已经圈定了 23 处应急供水地下水源地，以 2010 年为限，应急供水能力为 $10 \times 10^8 \mathrm{m}^3/\mathrm{a}$。

参 考 文 献

[1] 中国地质调查局. 中国地下水与资源调查成果报告[R]. 2005.
[2] 陈梦熊, 马凤山. 中国地下水资源与环境[M]. 北京：地震出版社, 2002.
[3] 刘昌明, 陈志恺. 中国水资源现状评价和供需发展趋势分析[C]//中国可持续发展水资源战略研究报告集. 第 2 卷. 北京：中国水利水电出版社, 2002.
[4] 潘家铮, 张泽祯. 中国北方地区水资源的合理配置和南水北调问题[C]//中国可持续发展水资源战略研究报告集. 第 8 卷. 北京：中国水利水电出版社, 2002.
[5] 段永侯, 肖国强. 河北平原地下水资源与可持续利用[J]. 水文地质工程地质, 2003, 30(1): 1-7.
[6] 北京市地质调查研究院. 首都地区地下水资源和环境调查评价报告[R]. 2003.

An Integrated Groundwater Management GIS to Improve Water Supply Safeguard for Emergency Well-field, Beijing

Wei Jiahua[1] Li Yu[2]

(1 Department of Hydraulic and Hydropower Engineering, Tsinghua University, Beijing, China 100084; 2 Beijing Institute of Geoengineering Exploration   Beijing, China 100037)

1 Introduction

Groundwater is a very important natural resource widely used for different purposes like drinking, irrigation, industrial use etc. in Beijing. With the development of economy and municipal construction, the water demand is continuously increasing, groundwater is becoming a natural resource of strategic importance. In order to solve the urgent problem, sometimes, there is no choice but to overexploit groundwater temporarily before South to North water division. To satisfy the increasing water demand and dealing with the dry season, an emergency well-field had been founded in 2001. Compared with surface reservoir, the groundwater reservoir not only has such advantages as large volume, high water quality and not easy to be polluted, but also being free of deposition and hidden trouble on security. It is, without question, a better engineering for the adjustment and storage of water resources.

The study area, is a watershed basin in north part of Beijing City. The investigated area is approximately 384 km^2 and consist of three districts: Miyun, Huairou and Shunyi. The climate is considered arid(semiarid) where the average amount of precipitation and evaporation are 613 mm/a, 1,600 mm/a respectively for a period of 1915~2000.There exists a great difference between the annual precipitations from 300 mm in extremely arid year to 1,400 mm in extremely heavy precipitation year, and inner-year, the precipitation concentrates in July and August, taking up 60% of the annual amount. The quite asymmetric precipitation can easily result in the surface runoff, which consequently causes the undesirable loss of water resources.

Three seasonal rivers run across this area, namely Chaobai River, Huai River and Yanxi River. In the surrounding area, there also exist three large surface reservoirs, namely Guanting Reservoir, Miyun Reservoir and Huairou Reservoir. These three reservoirs play an important role in storing the surface runoff during the rainy period. By regularly control the water discharge from these reservoirs to the basin in non-precipitation time, the groundwater reservoir can be recharged timely.

The aquifer system, its interaction with surface water, and its response to withdrawals, have been previously studied by the Beijing Institute of Geo-engineering Exploration (BIGE) (zhang Y. D., 2002). The subsurface comprises the Quaternary alluvium or pluvial materials, which is the primary aquifer in the area, and the thickness of the Quaternary sediment ranges from 180 m to 250 m. The hydraulic conductivity of the Quaternary sediment, comprising sand with clay and gravel lenses, ranged from 10 ~ 200 m/d across the domain.

Before 1979, the groundwater was in natural state and the groundwater level was mainly controlled by the atmospheric precipitation and the water level in the rivers. It fluctuated regularly within the year round. But the natural state was broken and the flow field was changed locally since the foundation of the Well-field in this area in 1980. Since 1983, the average production stabilizes at about 410,000 m^3/d. Until 1985, because of the influence of successive years of scarce precipitation, the water level continuously went down near the production area. For the purpose of stabilizing the groundwater level, which in turn will maintain the steady production, a regulation surface reservoir was constructed south of study area near the Xiangyang Village in 1985. From then on, the water level was maintained at about 32 m. Soon both the regional and local water level recovered rapidly. At the end of the year, the water level nearly reached the initial sates. To evaluate and management the existing and new well-field, the groundwater management information system is developed.

2 EGWMIS Development and Function

Development and management of groundwater resource requires a thorough understanding of the quantity and quality of groundwater aquifers. The groundwater system infrastructure and related data that were collected for the program consists of aquifer structure, water supply networks, precipitation, evaporation, water table and water quality, and so on. The EGWMIS has three major systems. The first is the data collection system, which collects the basic information such as data, image, and accurate transmission modes such as network based on TCP/IP. The second is the data storage and processing system, which stores and processes the mega data of the groundwater model and thematic analysis by powerful computer system with bulk memory and high-speed calculation capacity. The third is the decision support system, which simulates in real-time the groundwater system, optimizes the groundwater abstracting schemes, and assesses the quality of groundwater.

2.1 Methodology

The EGWMIS is developed with Microsoft Visual Studio 6.0 and ESRI Mapobjects 2.2. Mapobjects is based upon Microsoft's Component Object Module (COM) technology. This technology allows developers to expand ArcGIS platforms and develop customized applications. DLL's allow access to core GIS capabilities. With this new technology, it is possible to use GIS's data management capabilities. Microsoft Access is used to establish attribute database.

2.2 Background and Dynamic Data

Since water is predominantly extracted from aquifers through water wells, well records are the focus of both spatial and tabular data in GIS. Well records are kept by the BIGE. Well records include information regarding owner and driller, casing and pumping information, well location such as latitude and longitude, and types of logs associated with the well. Other information pertinent to the creation of management tools to assist a groundwater conservation district includes base map data, land use data, weather station data such as precipitation and evaporation records, and groundwater chemical data. Background data are used to evaluate water quality change. Well and observed bores record data from the BIGE contains historic water level measurements for aquifers and stream runoff. This data can be visualized by GIS for a range of years to show water level trends for a period. The dynamic of groundwater are observed every day.

A database was created in Microsoft Access containing all the information available in the well records. Such information includes the well and driller information, available logs, water level measurements, pumping and casing data, and well location. Also, in expectation of future study and analysis, additional areas of data input were created for water quality, water pumping. The database management interface permits the management to immediately update their observed groundwater dynamic information and plot thematic maps that reveal the changes of groundwater table.

2.3 GIS Spatial Data

Most of the EGWMIS functions are supported by the GIS data developed in this project. The first step in managing the groundwater resources of the EGWMIS is to identify the wells and observed bores within the affected area. Well records were acquired in both a GIS format and in a tabular format with latitude and longitude coordinates. Incorporating these wells with base map data allows the manager to visually locate all the wells, which enhances their analyses of well monitoring data such as water levels, water quality, and well yields. In addition, since the database also contains information regarding the well depth, aquifer code, well type, water use, and available logs, the manager can instantly identify many well parameters throughout the area.

The topography of the study area was created using the State Bureau of Surveying and Mapping 1:250,000 DLG data. Topography data is required to calculate an aquifer's physical thickness as well as its historic saturated thickness. The thickness for the aquifers was developed from driller's logs identified in the well records from the Beijing Institute of Geo-engineering. The geologic formations identified from the logs and their depths were entered into a database. This data table was joined by the well ID number to the well data shape files, which includes two layers: one layer representing the depth to the top of the aquifer and the other representing the depth to the aquifer bottom.

2.4 Groundwater simulation model

Most of general GIS can easily accomplish overlay and spatial analysis, but cannot perform the process-based groundwater modeling functions related to groundwater flow and transport processes. GIS has been widely adopted for use with groundwater models providing functions for data storage, calculation of required parameters, data manipulation, and output processing. GIS technology is also central to these models by providing the system with spatial data management, analysis, display and interface functions (Rifai, H. S., 1993; Richards, C. J., 1996; Sui, D.Z, 1999; WEI J. H., 2001). The groundwater numeric model, on the other hand, plays an important role in studying the movement of groundwater in porous media. By obtaining, operating and displaying the model dependent spatial data and the simulation results, the application of GIS to groundwater study can further detail the model and accelerate the realization of the occurrence and movement of groundwater in the aquifer. Coupling a GIS to groundwater numeric models can provide an efficient tool for processing, assessing, storing, manipulating, and displaying hydrogeological data.

In such a system it is often difficult to derive sensible data sets against which to test and calibrate such a model as this. Recharge may be directly measured but the data will be site specific and therefore difficult to relate to the larger scale. No such measurements are available for this study area though previous studies have estimated recharge using a numerical approach. The study by Wei, J. H. (2001) and Wang, G. Q. (2004) derived recharge estimates based on modeling of the groundwater flow system. The dynamic of groundwater table and volume are obtained from the GIS-based 3D groundwater flow model(equation (1)) of the aquifer that is currently under using. Figure 1 is the flow chart of groundwater analysis system based on GIS and groundwater numerical model.

$$\left.\begin{array}{ll}\mu\dfrac{\partial h}{\partial t}=\dfrac{\partial}{\partial x}\left(K_x\dfrac{\partial h}{\partial x}\right)+\dfrac{\partial}{\partial y}\left(K_y\dfrac{\partial h}{\partial y}\right)+\dfrac{\partial}{\partial z}\left(K_z\dfrac{\partial h}{\partial z}\right)+P & (x,y,z\in\Omega,t\geq 0) \\ h(x,y,z,t)\big|_{t=0}=h_0(x,y,0) & (x,y,z\in\Omega,t\geq 0) \\ h=h_1(x,y,t) & (x,y,z\in\Gamma_1,t\geq 0) \\ K_n\dfrac{\partial h}{\partial \vec{n}}\bigg|\Gamma_2=q(x,y,z,t) & (x,y,z\in\Gamma_2,t\geq 0)\end{array}\right\} \quad (1)$$

where, Ω is affected area. h is the water level, m. $h_1(x,y,t)$ is water level on known water level boundary. $h_0(x,y,0)$ is initial water level, m. k_x、k_y、k_z is x, y and z direction equivalent hydraulical conductivity. μ is specific yield. P is the vertical exchange volume. $q(x,y,z,t)$ is lateral recharge (discharge) per unit length per unit time, recharge use positive, discharge use negative, zero for impermeability boundary. Γ_1 is known water level boundary. Γ_2 is permeable and impermeability boundary.

The simulation model is used for determination of available groundwater storage. The desired result of these calculations is to be a visual description of the various layers in the area and a numerical estimate for the amount of available in the area, which is very useful for water early warning.

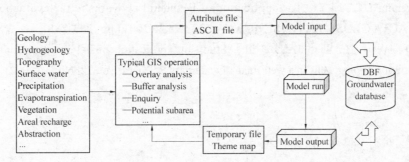

Figure 1 Flow-chart for GIS-based groundwater assessment

2.5 Groundwater Quality Assessment

The groundwater chemical components are assessed by single factor and multi-factor evaluation method. One factor beyond the limit, the system will give a suggestive marked at that point. The background and period of time quality map can be obtained by spatial analysis though polygonal layers derived from linear interpolation.

3 Conclusion

The management of groundwater dynamic data and evaluation of groundwater is a complex task. Traditional methods of groundwater resources computer modeling have both advantages and disadvantages. GIS software can better manage the data. But universal GIS software requires scripts and macros to perform water resource analysis and is more difficult to edit. The integration GIS, database and groundwater model are the ideal tools to assist groundwater management. The integrated system is not only usefulness for assessment groundwater by simulation model, but convenient for the decision maker. The system is continuing to achieve the benefits of EGWMIS as evident by comprehensive groundwater evaluations and well-field water quality assessment. EGWMIS is also continuing to help in the day-to-day decision-making process resulting in avoiding unnecessary trouble as early warning in aspect of water volume and quality. The next step work is to develop groundwater quality simulation model.

References

[1] Richards C. J., Roaza H. P. and Pratt T. R. . "Applying Geographic Information Systems to ground Water Assessments", Symposium on GIS and Water Resources (Poster Session), Fort Lauderdale, Florida, American Water Resources Association, September 1996.

[2] Rifai H. S., Hendricks L. A., Kilborn K. and Bedient P.B., "A geographic information system (GIS) user interface for delineating wellhead protection areas." Groundwater, Vol. 31, No. 3, 1993, pp:480-488.

[3] Sui D. Z., Maggio R. C.. "Integrating GIS with hydrological modeling: practices, problems, and prospects", Computers, Environment and Urban Systems, Vol. 23,(1999),pp: 33-51.

[4] Wang G. Q., Wei J. H., and Zhang Y. D.. "Analysis of the Sustainability of the Development of a Small Phreatic Aquifer in Northern China. Water International, Vol.29, No.4,(2004),pp: 467- 474.

[5] Wei J. H.. "Groundwater geographic information system: integration, visualization and case study" Dissertation submit to China University of Geosciences(Beijing)for doctoral degree, 2001.

[6] Wei J. H., Zhang J. L., Li Y.. "The application of BP neural networks to the evaluation of well field influence". ACTA GEOSCIENTIA SINICA, Vol.22, No.3,(2001),pp: 283-288.

[7] Zhang Y. D., Wei J. H., Shao J. L., et al.. "Study on application of 0-1 integer programming to the optimal layout of pumping wells in a well filed". Quaternary Sciences, Vol. 22, No. 2, (2002), pp: 141-147.

济南市玉绣河水生态环境修复综合措施

王 琳

(青岛海洋大学,山东,青岛 266003;山东省侨务办公室,山东,济南 250013)

1 玉绣河改造前状况

玉绣河是济南市区南部的一条城市河流,全长 10.5 km,高差约 136 m,流域面积 73.58 km²。由于城市化的发展,河流污染严重。为配合南水北调工程,在对玉绣河生态现状分析的基础上,济南市园林局从点源污染治理、开放式渗渠构造、河滨带景观生态建设、生态型护岸、优化河床结构、景观塘建设等六方面着手,开展玉绣河的改造治理和生态修复工程。

2 景观生态建设(一期工程)

2.1 工程措施

目前玉绣河生态治理一期工程已进行完毕,完成投资额 6 000 万元。具体措施包括以下方面:

(1)底泥疏竣截污。对现有河道内的非点源性污染源底泥进行疏浚,在点源污水较集中的河道旁分散建设中水站,对点源污水进行深度处理。

(2)水体复氧。沿玉绣河设置 5 座跌水堰,利用水坝的跌水和河道纵断面的近自然处理等进行曝气增氧。其他河段沿主河道每间隔约 200 m 修建了滚水坝、橡皮坝等进行蓄水,增加渗漏,补充泉源。

(3)生态化措施。玉绣河植物园附近建设水面面积约 2 200 m² 的景观塘,提高水体景观性和水域净化能力,改变水环境生态链结构的单一性。

(4)生态河道断面设计。采用多种河道断面形式,改变水体状态和增减水岸遮蔽物等方式,降低水流速度,为河流水生生物提供栖息场所,增加生境多样性和物种多样性,形成稳定的河流生态系统。

(5)河滨带治理。拆除沿岸违章建筑和部分建筑,玉绣河原有明沟两侧,每侧留 15 m 宽的景观植被带,在沿途人类活动较集中的地段做节点。在景观生态节点处结合蓄水,建设小游园,营造不同形式的城市水景和园林小品,为市民提供休闲、锻炼的空间。

通过以上措施,大大增加了玉绣河的河道自净功能,玉绣河已成为一条贯穿济南市城区南北的绿色生态廊道。

2.2 效果评价

2.2.1 植被状况

通过对比改造前后玉绣河两岸的景观物种多样性指数 H、景观物种优势度指数 D 和景观物种均匀度指标 E,玉绣河两岸的植被生态已达到较好的效果。

景观物种多样性指数方面,舜耕广场>济大路—舜玉路>舜玉路—八里洼路,其平均值分别为 1.502、1.466、1.392;景观物种优势度指数方面,三段的大小关系正好和上述多样性指数相反,舜玉路—八里洼路>济大路—舜玉路>舜耕广场,平均值分别为 0.503、0.429、0.393,舜玉路—八里洼路地段的优势度最大,此地段内物种繁多复杂且差异较大,生态结构稳定;景观物种均匀度在所调查的 18 个样方中有 6 个属于优良级、11 个属于良好级,只有 1 个属于差级。

2.2.2 水质变化

改造后，玉绣河的河流水质有了很大的改善(见表1)。

表1 玉绣河改造前后水质对比

序号	指标	单位	改造前	改造后
1	COD_{Cr}	mg/L	400	105.99
2	NH_3-N	mg/L	40	4.89
3	TN(以N计)	mg/L	50	12.56
4	TP(以P计)	mg/L	3.5	0.98

注：改造前的水质引自《济南市玉绣河综合治理引水工程污水站招标文件》，改造后的水质为现场监测平均值。

3 河流生态系统健康现状评价

运用层次综合法评价了玉绣河广场东沟和广场西沟的健康状况。广场东沟的单因子指标健康状况相对较差，其主要限制因素是水质恶化、水生物多样性的消失以及河岸带被挤占。因此，对广场东沟应重点从优化河流水质、修复水生物栖息环境以及恢复河滨带景观生态方面整治。而对广场西沟河道整治与管理中，消除点源污染、控制面源污染、修复水生物栖息环境等是治理重点。广场东沟的综合评价值为1.13，处于亚健康状态；广场西沟的综合评价值为2.16，处于健康状态。对其进行等权处理，玉绣河整体的综合评价值为1.65，处于亚健康状态。

4 河流治理途径

4.1 出水水质控制指标选择

本研究采用灰关联法分析玉绣河水体中叶绿素a与其他指标间的关联关系，通过实验室静态和动态模拟小试验确定水体浮游藻类的限制性营养因子，作为玉绣河点源污染分散处理单元出水水质的重点控制指标和计算稀释生态需水量的依据。

4.1.1 藻类与各营养因子的关联分析

(1)样品采集。采样点选择玉绣河植物园的景观塘下游处，监测时间在5~11月期间，每月连续7天，每天上午和下午两次采样分析，取其平均值作为每月的水质平均浓度；遇降雨事件时暂停取样，待河流流量恢复至降雨前水平时继续取样。

(2)样品分析。采样现场采用便携式设备检测水体的温度(Tem)和溶解氧(DO)。样品在取回实验室的当天，检测COD、TN、NO_3^--N、NH_3-N、NH_3-N、NH_3-N、TP、PO_4^{3-}和叶绿素a(Chl-a)的浓度。

(3)水质数据。2006年5~11月的水质监测结果汇总见表2。

表2 5~11月期间玉绣河水体理化指标汇总表

月份	Chl-a	TN	NO_3^--N	NH_3-N	COD	NO_2^--N	TP	PO_4^{3-}	DO	Tem
5	8.3	16.56	0.41	7.32	60.75	0.52	1.02	0.72	5.54	20.77
6	9.14	13.74	0.4	6.2	91.22	0.6	1.04	0.79	6.51	24.23
7	43.94	9.44	1.41	3.85	83.9	0.6	0.87	0.6	5.47	27.8
8	40.68	9.39	1.08	3.03	123.27	0.39	0.81	0.6	5.22	27.43
9	49.36	10.49	1.79	7.06	82.27	0.44	1.01	0.75	6.03	22.26
10	35.7	13.59	2.24	7.18	73.49	0.43	1.23	0.83	6.34	20.63
11	28.09	4.86	1.84	2.34	65.36	0.34	1.05	0.62	8.06	8.83

注：Chl-a的单位为mg/m^3，Tem的单位为℃，其他指标的单位均为mg/L。

表2数据显示玉绣河的水质已劣于地表水V类水质，水体呈富营养化状态。经现场调查，玉

绣河的污染物主要来自上游未改造村庄的点源污染、上游农业区以及土壤侵蚀等面源污染。

(4)关联分析。本研究采用灰关联法分析玉绣河景观水体中叶绿素 a 与其他 9 项理化指标间的关联关系。计算方法是对数据进行均值化的无量纲化处理，然后对灰关联系数 $\varepsilon_i(k)$ 和灰关联度 γ_i 进行计算。

4.1.2 藻类的限制性营养因子分析

灰关联分析结果表明，5~11 月各指标与 Chl-a 的关联度相似，与 Chl-a 浓度关联度较大的指标依次是 NO_3^--N、Tem；但 5~10 月 NO_3^--N 与 Chl-a 的关联度（$\gamma=0.74$）明显高于其他，5~9 月的这种现象更为明显（$\gamma=0.88$），这与刘冬燕等分析苏州河水质后得出 NO_3^--N 含量与 Chl-a 呈明显的正相关关系的结论相一致。但在 11 月份 NO_3^--N 与 Chl-a 的关联度降低，从表 2 可以看出，11 月份的水体温度明显低于其他月份，温度过低会导致生物活性降低，表现为 Chl-a 浓度下降，此时温度对浮游藻类的抑制性作用增加。因此，通过控制限制性营养因子的方法对春夏季节的浮游藻类进行有效的抑制后，秋冬季节的浮游藻类也同样会得到有效的抑制。

我们还可以看出，NH_3-N 与 Chl-a 有着较差的关联关系，在 5~9 月、5~10 月的数据分析中，其关联度均表现为最差，但监测结果(表 2)表明水体中 NH_3-N 的平均浓度约为 NO_3^--N 浓度的 4 倍，这与普遍认为的"藻类优先利用 NH_3-N，而且 NH_3-N 的存在还会抑制对 NO_3^--N 的吸收"相矛盾，因此分析认为藻类对 NH_3-N 的吸收存在较为特殊的内在特征。

4.1.3 静态条件下藻类对限制性营养因子的吸收

(1)试验设计。取 15 L 玉绣河景观塘河水置于水族缸中，放置在向阳处，采用 90 r/min 的搅拌器进行慢速搅拌，以促进水体复氧和防止藻类下沉，消除器壁效应。要求试验水体温度在 8.3~17.3 ℃间波动，溶解氧为 4.3~8.7 mg/L，基本与玉绣河水体条件相似。每天上午 10:00 取样，测试试验水体中 NO_3^--N、NO_2^--N、NH_4^+-N、PO_4^{3-} 和叶绿素 a(Chl-a)的浓度，以及溶解氧和温度值，连续监测 20 天。

(2)试验结果与讨论。实验结果显示，氮素是玉绣河水体中藻类的限制性营养因子，在 NH_4^+-N 和 NO_3^--N 的浓度比达到 8:1 时，藻类优先吸收利用 NH_4^+-N；由于发生亚硝化和硝化反应，导致 NO_3^--N 与藻类生长呈正相关。因此，对于玉绣河水体，NH_4^+-N 是浮游藻类的限制性营养因子，如果通过某种措施降低水体中 NH_4^+-N 的浓度，水体中藻类会因失去氮素的营养支持而消亡，以此达到控制水体中藻类的目的。

4.2 分散处理单元规模

4.2.1 一维水质模型

绝大多数城市河流的水质计算常常可以简化为一维水质问题，即假定污染浓度在河流横断面上均匀一致，只沿纵向流程方向发生变化，则河流水质迁移转化的基本方程可简化为一维方程；假定河段流量和排污稳定，各断面的污染物浓度不随时间变化，忽略河流纵向弥散作用，则可得稳态条件下的一维水质迁移转化方程：

$$c = c_0 \exp\left(-\frac{kl}{u_x}\right) \tag{1}$$

式中：c 为河流景观水体拟控制的限制性营养因子浓度，mg/L；c_0 为上游分散处理单元所在河流断面的水质要求，mg/L；k 为河流的污染物迁移转化系数；u_x 为引水后河流流速，m/s；l 为出水断面与控制断面间的距离，m。

本课题研究以广场西沟为研究对象，中水站 1 和中水站 2 间的河段没有点源污染输入，可将其视为运用一维水质模型的理想河段。该河段长约 2 000 m，流速为 0.09 m/s，流量为 0.14 m³/s。选取广场西沟上舜玉路断面和植物园景观塘入口断面为监测取样点，根据 2006 年 11 月下旬和 2007 年 1 月下旬监测的数据，利用 Excel 的数据分析功能进行回归分析，建立回归模型：

$c=0.7512c_0$ ($n=9$, $r=0.9636$, $p<0.01$),表明玉绣河 c 和 c_0 在 1%的置信水平下是显著相关的,回归模型成立。

计算回归函数,解得 $k=1.287\mathrm{e}^{-5}$,则玉绣河的一维水质迁移转化模型为:

$$c = c_0 \exp\left(-\frac{1.287\mathrm{e}^{-5} l}{u_x}\right) \tag{2}$$

4.2.2 河流稀释净化需水量计算

本课题以污染物稀释净化需水量作为改善城市河流生态环境所需的水量,采用河段特定功能区控制法计算稀释净化需水量。在河流流量和污染物浓度相对稳定的前提下,在分散处理单元出水口(k 断面)输出水量 Q_k、污染物浓度 C_k,经过河段 l 的迁移转化降解,到达功能水体入口断面 m 时,污染物的浓度达到该处水质约束指标的目标值 C_m。依此反推,求得河道稀释污染物所需的引水量。

满足玉绣河稀释污染物浓度达到功能区水质目标的引水流量 Q_k 和 Q_w 计算方程为:

$$(aQ_k + Q_w)C_k - 2(Q_k + Q_w) = 0.28 - 0.988a \tag{3}$$

4.3 分散处理模式应用的经济分析

4.3.1 规模效益计算

某一河段的点源污染集中处理可以减少污水处理厂的座数,节省基建总投资。龙腾锐等推导出污水分散处理与集中处理的污水处理厂所需投入基建费用之差可用公式(4)表示:

$$E_m = \alpha \left(\sum_{i=1}^{N} Q_i^{\beta}\right)\left[1 - N^{(\beta-1)}\right] \tag{4}$$

式中:E_m 为规模效益,万元;Q_i 为分散处理单元规模,万 m^3/d;N 为分散处理单元个数;α、β 为系数,其中 $0<\beta<1$。

引用田一梅等参考国内常用的污水处理工艺,选择污水二级生物处理加深度处理的工艺流程,采用混凝沉淀→过滤→消毒的深度处理工艺而确立的费用函数模型系数,α 取 375.24,β 取 0.86。代入式(4)得:

$$E_m = 375.24 \left(\sum_{i=1}^{N} Q_i^{0.86}\right)\left[1 - N^{-0.14}\right] \tag{5}$$

4.3.2 中水输送费用

对于给定河段,中水回补河流的输送费用可表示为输水管线基建费用和中途泵站基建费用之和,即公式(6):

$$S_L = f(D)L + \sum_{i=1}^{m} S(Q_i, H_i) \tag{6}$$

式中:S_L 为距离为 L 时的总输送费用,万元;D 为输水管径,mm;$f(D)$ 为管径为 D 时,单位管线长度的基建投资,万元/m,该函数形式与管道取材以及当地的水文地质、气候等自然条件有关;L 为输送距离,m;m 为中水输送过程中需设置的中途泵站数;$S(Q_i, H_i)$ 为第 i 座泵站的基建费用,万元,本研究中 $S(Q_i, H_i)$ 采用文献提供的函数形式:$S(Q_i, H_i) = k_1(Q_i H_i)^{k_2}$,其中 k_1、k_2 为由实际数据拟合得到的费用系数。

通过对济南市多条不同管径管线实际造价以及多座不同流量扬程泵站实际造价的调查,进行回归分析和曲线拟合,得出济南市 $f(D)$ 的函数形式和泵站的费用系数 k_1、k_2。

4.4 分散处理单元的最佳分散度

4.4.1 选址原则

河滨带具有一定的空间,分散处理设施主体结构宜建于地下,最大程度地减小对周围环境和居住的影响；宜建在河流点源污染较集中的地方，以减少管网输送成本；尽量在河流上游选址；具备便利的交通条件。

4.4.2 最佳分散度定义

本研究采用最佳分散度的概念来计算确定某长度河道建设分散处理单元的规模，首先设定前提条件：在降雨充分时，分散处理单元出水用于满足河流污染物稀释净化需水；在干旱季节，分散处理单元出水全部用于满足附近绿地浇洒、植被蒸腾、河流水面蒸发和渗漏等城市环境需水；假定济南全市范围内绿化均匀；在上述选址原则的基础上应用最佳分散。根据上述设定，最佳分散度可表示为：

$$D=Q=f(2r) \quad 或 \quad D=Q_1+Q_2=f(2r_1+2r_2) \tag{7}$$

式中：D 为最佳分散度，m^3/d；Q_1、Q_2 为处理规模，m^3/d；r_1、r_2 为中水服务半径，km。

若沿河段分散处理单元个数增多，则式(7)可表示为通式：

$$D=\frac{\sum_{i=1}^{N}Q_i}{2\sum_{i=1}^{N}r_i} \tag{8}$$

4.4.3 环境需水量

城市环境需水量应主要包括城市绿地需水和城市河湖生态环境需水两部分。绿地需水量包括绿地植被蒸散需水量、植被生长需水量。城市绿地有园林、道路绿化带、河岸生态林、风景区林地等。此处所指的城市河湖生态环境需水量是指水面蒸发需水和渗漏需水。

根据济南市统计年鉴，2004年济南市绿地面积2163万 m^2，覆盖率39.1%。

(1)绿地植被蒸散需水量 W_{E1}。针对济南市的情况，取 360 mm/a。则：

$$W_{E1}=A_1E_P \tag{9}$$

式中：A_1 为绿地面积，km^2。

(2)绿地植被生长需水量 W_P。根据文献，按公式(10)计算绿地植被生长需水量：

$$W_P=W_{E1}/99 \tag{10}$$

(3)河流水面蒸发需水量 W_{E2}。根据杨志峰等的经验，年降水量在 500~980 mm，则年水面平均蒸发量 E_q 在 700~1100 mm，本研究取 1100 mm/a，则河流水面蒸发需水量可表示为：

$$W_{E2}=A_2E_q \tag{11}$$

式中：A_2 为河流水面面积，km^2。

(4)河流渗漏需水量 W_L。

根据文献，河流渗漏需水量 W_L 可表示为：

$$W_L=KA_2 \tag{12}$$

式中：K 取为 750 mm/a。

(5)总的环境需水量 W。总的环境需水量为各单项环境需水量之和，即：

$$W=W_{E1}+W_P+W_{E2}+W_L \tag{13}$$

4.4.4 最佳分散度计算

以规模为 Q 的单一处理单元为例，以其为圆心，服务半径为 r 范围内的绿地面积 $A_1=0.391\pi r^2$，附近河流水面面积 $A_2=2rb$（b 为河宽，玉绣河宽为 9 m）。将式(9)~式(12)以及 A_1、A_2 的表达式代入式(13)得到玉绣河附近沿河段长度为 $R=2r$ 范围内的城市环境需水量为：

$$W=0.1224r^2+0.0091r$$
$$W=0.0306R^2+0.0046R$$

式中，W 的单位为万 m^3/d。

根据最佳分散度的定义，$W=Q$，进行单位换算后得：

$$D=Q=306R^2+46R \tag{14}$$

即最佳分散度为河段长度的一元二次函数，当河段长度确定时，分散处理单元规模也就相应确定；相反，当分散处理单元规模确定时，其间隔距离也就相应确定，有利于分散处理单元的布局规划。

同理可以得到两座分散处理单元情况下，玉绣河的最佳分散度：

$$D = Q_1 + Q_2 = 306\left(R_1^2 + R_2^2\right) + 46\left(R_1 + R_2\right) \tag{15}$$

参 考 文 献

[1] 李玉华,郭立杰.哈尔滨市何家沟综合治理措施[J].哈尔滨工业大学学报，2004, 136 (14):536-538.

[2] 高桥幸彦,杜茂安, 等.污水处理排放水对小流量河流水体生态的影响[J]. 哈尔滨工业大学学报，2006, 38(2): 212-215.

[3] Narcis Part, Antoni Munne. Water use and stream flow in a Mediterranean stream [J]. Wat. Res, 2000, 34 (15): 3876-3881.

[4] 黄祥飞.湖泊生态调查观测与分析[M]. 北京:中国标准出版社,1999.

[5] 国家环境保护总局.水和废水监测分析方法[Z]. 4 版.北京: 中国环境科学出版社, 2002.

[6] 刘冬燕, 宋永昌. 苏州河叶绿素 a 动态特征及其与环境因子的关联分析[J]. 上海环境科学, 2003, 2(3): 261-264.

[7] Mc Carthy J J, Wynne D, et al. The uptake of dissolved nitrogenous nutrients by Lake Kinnert (Israel) microplankton [J]. Limnol. Oceanogr., 1982, 27: 673-680.

[8] Dorteh Q. The interaction between ammonium and nitrate uptake in phytoplankton [J]. Marine Ecology Progress Series. 1990, 61: 183-201.

[9] 龙腾锐, 郭劲松. 临界距离优化城市污水处理系统研究[J]. 中国给水排水, 1998, 14 (2):16-18.

[10] 田一梅, 赵新华, 等. 城市自来水与中水系统综合规划的优化研究[J]. 给水排水, 2001, 27 (5): 23-26.

[11] 赵元, 单金林. 城市给水输配系统加压泵站的优化计算[J]. 中国给水排水, 1999, 15(9): 31-33.

[12] 王敏, 郑新奇. 济南市生态环境需水量的计算[J]. 水利科技与经济, 2006, 12 (3):141-146.

[13] 杨志峰, 崔保山, 等. 生态环境需水量理论、方法与实践[M]. 北京: 科学出版社, 2003.

济南市水资源可持续利用实践与探索

孟庆斌

(济南市水利局，山东，济南 250001)

1 济南市水资源概况

1.1 水资源量

济南市水资源主要来自大气降水和过境河流两大部分，由当地地表水、地下水和客水组成。当地地表水主要由玉符河等河道及卧虎山水库等工程提供，地下水主要由裂隙岩溶水和第四系孔隙水构成，客水主要包括黄河、徒骇河、德惠新河等过境河道来水，南水北调东线工程建成后，长江水也将进入济南，成为新的客水资源。根据计算，济南市多年平均降水量为638mm，折合水量52.1亿 m^3，多年平均水资源量为19.59亿 m^3，其中地表水资源量7.47亿 m^3，地下水资源量12.12亿 m^3。全市人均水资源量351 m^3，仅为全国人均水资源量的1/6。

1.2 水资源特点

(1)水资源补给年内、年际变化大。汛期6～9月份降雨占全年雨量的70%，是水资源的主要形成期，春冬很少。7～8月份是河川径流的主要形成期。丰水年水资源量为枯水年的2～4倍。地表径流量丰水年与枯水年要相差几倍、十几倍，有的县在特枯年无地表径流量。降水量和径流量的年际变化大和年内高度集中造成了水旱灾害和水资源开发利用的困难。

(2)水资源地域分布不均。山丘区的地下水可利用量很少，多集中在山前平原一带；无拦蓄工程的地表水无可利用量而言，在北部平原商河县内有近1/2的面积无浅层淡水。

(3)水资源结构以地下水为主，泉水是济南水资源的特色。济南市水资源的组成以地下水为主，地表水与地下水相互转化迅速。济南地区有丰富的泉水资源，但随着用水量的增加，泉域地下水超采，泉水逐步由常年喷涌变为季节性、间歇性喷涌，甚至全年停喷。保持名泉长年持续喷涌和市民饮用优质地下水成为目前济南水资源管理的重要目标。

(4)人均水资源占有量少，且水资源开发利用结构不合理。全市人均水资源量351 m^3，地下水开发利用强度达85%。部分地区地下水严重超采，导致泉群停涌、地下水环境恶化及地裂地陷等一系列环境问题。

1.3 水资源开发利用情况

目前，全市已建成大中型水库12座、小型水库175座、塘坝900余座，总拦蓄能力5.2亿 m^3；建成邢家渡、田山等大中型灌区6处；建成配套机井5.4万眼。2006年全市供水总量15.46亿 m^3，其中当地地表水1.8亿 m^3，占总供水量的11.64%，地下水7.5亿 m^3，占总供水量的48.51%，引用黄河水6亿 m^3，占总供水量的38.81%，污水处理及雨水利用量0.16亿 m^3，占总供水量的1.04%。从供水结构分析，基本上地表水与地下水各占50%左右。

1.4 水质状况

(1)水库水。大多数水库水质符合Ⅰ～Ⅱ级标准，可作为饮用、灌溉、工业用水，担负城市供水、农业灌溉和生态用水功能的卧虎山水库蓄水量多时为Ⅱ级，枯水季节COD、BOD、总磷、总氮超标。

(2)黄河水。基本符合饮用水源水标准，个别时段砷、大肠杆菌、高锰酸钾、总汞等指标超标。

(3)小清河。除源头睦里庄断面符合国家地表Ⅱ类水质标准外，其余断面均为劣Ⅴ类水质，溶解氧、COD、高锰酸钾指数、BOD和氨氮指标超标，水体特征污染物为氨氮。

(4)地下水。南部山丘及山前平原区，大部分岩溶水和第四系潜水水质良好，城市局部地区出现污染问题，已检测出油类、六价铬、酚、汞、铅等多种有机污染物，地下水有加重的趋势。

2 水资源管理中的主要问题

2.1 缺水和浪费并存

根据全市水资源供需分析结果：2020年，保证率50%、75%、95%时，全市缺水率分别为18.1%、25.6%及28.1%，济南缺水严重。

另一方面，水资源利用率不高。据统计，2006年全市万元GDP取水量73.59 m^3，工业用水重复利用率为74%，农业灌溉水的有效利用率只有45%左右。群众节水意识较差，生活节水器具推广缓慢，用水浪费严重，同时，由于设施老化和缺乏行之有效的管理，城市公共供水管网漏失率高达30%。

2.2 水资源工程配置体系不完善

投资20多亿元的鹊山和玉清湖两大供水工程设计供水能力80万 m^3/d，目前实际供水量还不到设计供水能力的一半，且黄河水供水管网不能覆盖东部地区。而东部地区工业相对密集，用水量庞大，主要以地下水为供水水源。东部地区地下水的长期集中开采，形成超采区面积达200多 km^2，漏斗中心地下水埋深达20 m，成为影响生态环境的重要因素。

另外，中水资源、汛期雨洪资源也由于工程设施不健全而不能得到有效利用。

2.3 地下水涵养能力下降

南部山区是济南市区泉水的直接补给区，受地质条件约束，灰岩区土壤稀薄，林木植被稀少，地表水资源缺乏，生态环境脆弱。城市及乡镇的扩张，硬化面积加大，使地表径流增加、地下水入渗补给量减少。同时，对采石、挖土、无序开发等行为的管理不利，致使南部山区生态环境遭到一定程度的破坏。塘坝淤积、河道萎缩，也使南部山区的水源涵养能力明显下降。

2.4 水生态环境恶化状况仍未有效遏止

地下水的超量开采，导致地下水采补失衡，泉水不能持续喷涌。20世纪70年代前，济南市区泉群正常降水年份长年喷涌，多年平均日涌水量35万 m^3。自1972年开始，泉水相继出现短暂停喷和季节性断流。90年代停喷加剧，并出现了年际间断流，最长的停喷发生在1999年3月～2001年9月，断流时间长达926 d。

3 济南市水资源管理实践及探索

济南水资源管理的近期主要目标是：实现正常降水年份泉水常年喷涌，实现城市居民饮用优质地下水，确保济南未来发展不缺水，以水资源的可持续利用支撑经济、社会科学和谐发展。水资源开发利用遵循的基本原则：充分利用地表水，合理开采地下水，积极引蓄黄河水，大力推行节约用水。

3.1 改革水资源管理体制

济南市的水资源管理经历了由城乡分割、部门分割向统一管理的过程。2000年8月，市水利局对全市水资源实行统一管理。2001年9月20日济南市颁布实施了《济南市水资源管理办法》，为水资源管理提供了法律保障。2005年5月将原隶属于市政公用事业局的玉清湖、鹊山水库水源地和地下水井泵整建制移交市水利局，实行水资源统一管理。2006年3月1日，城市公共供水原水资产全部移交水利局。至此，济南市水资源管理工作基本实现了真正意义的统一管理。

3.2 积极构筑水资源工程保障体系

为了适应济南市新一轮城市规划建设对水资源保障提出的新要求，市领导提出了加强水资源能力建设的新规划：一是"南水北调"东线联结工程，将济平干渠与市田山引黄工程、玉清湖供水工程等现有骨干工程进行联结，进一步完善骨干供水网络；二是实施东联供水工程，利用现有的鹊山水库、章丘市的杜张水库和朱各务水库作为调蓄水库，新铺设输水管道，联合调度黄河水、地表水和明水泉水，向东部工业区和东部产业带内的济南钢铁集团、章丘电厂、济南炼油厂等大型企业供水，改变市东部工业区以地下水为主要水源的局面，实现地表水置换地下水，促进优水优用和泉水保护。

3.3 实施封井保泉措施

对自备井限采或禁采是保护泉水、实现水资源可持续利用的重要举措。自2001年开始，济南市对二环路内具备条件的单位实施了自备井封闭工程。截至目前，全市累计封闭公共供水水源井以及单位深层自备井达到307眼、浅层水井1700余眼，年减少地下水开采8000多万 m^3。同时，坚决禁止在供水管网覆盖范围内开凿自备水井，城市规划新区内限制开采地下水。

3.4 建设节水型社会

通过节水宣传进社区、进校园活动，广泛开展节水宣传，提高全社会节约水、珍惜水、爱护水的意识，促进了节水器具进万家工作的开展。在工业节水工作中，在全市大中型企业中组织开展了水平衡测试，积极提供节水技术、信息咨询，推广使用先进节水工艺和设施，逐步建立健全了节水控制指标体系，使全市工业用水重复利用率提高到74%，万元GDP取水量减少到73.59 m^3，节水降耗指标达到全省领先水平。在农业节水工作方面，积极推广节水灌溉，全市节水灌溉面积达到8.67万 hm^2，占全市有效灌溉面积的37%。

发挥水价在节水中的经济杠杆作用，分别于2003年和2005年，全市全面提高水资源费标准。

3.5 建设水资源监测系统

2001年建成全市地下水位遥测系统，该系统由18个地下水位自动监测点构成，能够实时监测市区地下水位动态变化，在供水保泉过程中发挥了非常重要的作用。目前正在实施系统扩建工程，在东郊及市区南部增设27个自动监测点，使系统更加完善。

3.6 加强水源涵养与保护

将南部山区定位于"南控"区，划定红线，成立保护机构；以大流域为骨干、小流域为单元开展水土保护综合治理；在南山区实施了水库除险加固增容工程和河道拦蓄治理工程，增加南部山区山洪拦蓄能力；对宜林荒山进行全面绿化；实施人工增雨工程，在南部山区布设了110个人工影响天气站点，一遇适宜气候条件立即实施人工增雨作业。

3.7 开展保泉试验研究

玉符河等泉水补给强渗漏带实施回灌补源工程。2001年实施了第一次玉符河回灌补源试验，之后济南市先后进行了4次回灌补源，取得良好的效果。回灌补源优点明显：一是经常性补给地下水，对于保持泉城特色，实现水循环发挥重要作用；二是将地表水转化为优质地下水资源，使水质通过自然系统得到净化，成为一种经济的水质净化、水环境保护途径；三是将地表水转化为地下水，储存在地下水库中。

实施济西抽水试验。2003年1月，济南市进行应急供水与抽水试验，先后建成桥子李、冷庄和古城水源地，加上原有的峨眉山和大杨庄水源地，于6月开始启动应急供水与抽水试验，每天从上述水源地抽取30万～35万 m^3 地下水通过输水管道进入济南市供水系统，历时50d。2003年、2004年、2005年济南市连续三年丰水年，降水充沛，年降雨量分别为899 mm、891 mm和738 mm。2004年7月，济西抽水试验重新启动，至2005年12月，济西抽水试验持续进行1年零5个月，开采量30万 m^3/d，试验开采量累计达1.5亿 m^3。

济南市水的问题非常复杂，特别是对地下水的研究分歧较大。对以上两个试验，也存有不同的意见。

对玉符河回灌补源试验结论的不同意见主要是：补充到地下的水没有补到市区泉域，而是补到了济西。

对济西试验的分歧，主要是泉域边界问题，即济西地下水与市区泉群的关系问题。一部分专家认为，济南泉域边界东到东梧断裂，西到马山断裂，南到齐长城岭，北到济南火成岩体，面积 1 500 km²，济西地下水系统是济南泉域地下水系统的有机组成部分，两者具有密切的水力联系，泉群断流是东郊、西郊及城区大量开采地下水所致，开采济西地下水会对市区泉群造成影响，济西地下水开采量多年平均条件下应控制在 20 万 m³/d 以下。也有部分专家认为，济南泉域边界东到埠村向斜构造，西至郎茂山—万灵山岩溶弱发育带，南达锦绣川至以北的区域分水岭，北到济南火成岩体的北东向阻水前沿，面积 818.5 km²。泉水断流的原因主要是东郊地区和城区过量开采地下水。以隔水的济南火成岩体及与之相连的郎茂山—万灵山岩溶弱发育带为界，济西地下水为独立于济南泉域之外的地下水系统，两者不存在水量交换，开采济西地下水对济南泉水不会产生影响，其允许开采量为 60 万 m³/d。

泉域边界观点之争论结果导致在保泉指导思想和具体措施的不一致，究竟是"采西停东"还是"东西限采"，东西部地下水和泉域地下水究竟是"一碗水"还是"两碗水"，这期间曾进行过各种试验研究，但似乎还是不能达成共识，影响保泉决策与措施的有效实施和发挥作用。

山东省黄泛平原深层地下水资源可持续利用

徐军祥

(山东省地矿局，山东，济南 250011)

山东省黄泛平原属北方缺水地区，区内人均水资源占有量仅 460 m³/a，人均水资源占有量约为全国人均的 1/5，同时存在空间上分布不均。随着经济社会的发展，水资源短缺已成为制约该地区可持续发展的重要因素。由于深层地下水的超采，部分城市周围已经形成了深层地下水降落漏斗，由此产生地面沉降。据测量资料分析，随着德州市深层地下水开采量的增加，2000 年沉降区面积已达 2 037.5 km²，德州城区累计沉降量 150~387 mm，多年平均沉降量 22.0 mm/a。

1 深层地下水开发利用价值

(1)深层地下水的概念。本文所指深层地下水即深层的具有承压性的孔隙淡水，埋藏深度暂定为 200~800 m。

(2)山东省深层地下水资源分布。按照开采资源模数，山东省深层地下水资源大致可以分为三个区，即 $<0.75 \times 10^4$ m³/(a·km²)、$(0.75 \sim 1.0) \times 10^4$ m³/(a·km²)、$(1.0 \sim 2.0) \times 10^4$ m³/(a·km²)。

(3)深层地下水开发利用价值。目前深层地下水仍作为鲁西北、鲁西南地区城市供水的水源之一。如德州、滨州、济宁、菏泽等地级市一直把深层地下水作为城市供水的重要水源，高唐、临清、临邑、惠民等县城也是如此。

深层地下水可作为部分地区浅层水水质不佳的替代水源。如德州、滨州、菏泽市及下属部分县市的居民生活用水。同时，还可作为应急情况下的生活供水水源。如 1997 年，黄河断流 226 d，鲁西、鲁北、鲁西南的大部分地区出现人畜供水困难，德州市、滨州、济宁、菏泽市等主要城镇地区均利用深层地下水进行应急供水。

2 深层地下水资源的组成

深层地下水资源组成如下：

(1)越流补给。是指含水层通过相邻含水层的越流作用而得到的补给。在部分地区，深层地下水可接受上层含水层的越流补给。越流补给量占深层地下水资源的 50%左右。

(2)径流补给。深层地下水的补给条件很差，天然状态下，其补给主要来自上游(鲁中山区、太行山脉)地下水的侧向径流补给，补给区远，水交替微弱，径流极其缓慢，补给量很小。侧向径流补给占深层地下水资源的 15%左右。

(3)压缩释水。在开采条件下，含水层中的水被抽出，黏性土层被压密，由此要释放一部分水补给含水层，此水是山东省内深层地下水可采资源量的重要组成部分。压缩释水量占深层地下水资源的 30%左右。

(4)弹性释水。在承压含水层中随着水头的降低引起水的膨胀而释放一部分水。山东省深层地下水的水头一般高于含水层顶板 200~500 m，是可采资源量的重要组成部分。弹性释水量占深层地下水资源的 5%左右。

(5)储存量。指含水层中储存的重力水的总体积，相当于静储量。实际开采中，很难开采到这部分资源量。

就山东省黄泛平原区而言，深层地下水的径流补给是十分缓慢的(特别是在小清河以北和湖西地区)，在计算可采资源量时，径流补给量和储存量可不予计算，一般只考虑越流补给、弹性释水、压缩释水这三部分资源量。

3 深层地下水的水文地质特征

3.1 深层含水层组及富水性

深层地下水分为中深层淡水与深层淡水，中深层淡水由于水量较小不单独进行开采，而是与深层淡水作为一个统一含水层进行开采。按照顶板埋深的不同大致可以分为4个区，即<100 m、100~200 m、200~300 m、300~400 m，含水层顶板埋深的分布在水平方向上具有一定的规律性，即靠近山前的地带埋深较浅，向平原区腹地埋藏深度逐渐增大。按照富水性的不同，大致可以分为3个级别，即<500 m^3/d、500~1 000 m^3/d、1 000~3 000 m^3/d。富水性的分布基本与砂层厚度的分布密切相关，一般砂层厚度大的地段富水性较好。

深层地下水富水性在水平方向上具有较为明显的分带性，即山前水量较小，向平原腹地水量有逐渐增大的趋势。

3.2 深层地下水补给、径流和排泄条件

(1)补给条件。深层地下水的补给条件很差，天然状态下，其补给主要来自上游地下水的侧向径流补给，补给量很小。开采状态下，除接受侧向径流补给外，还接受上覆含水层越流补给和黏性土压缩释水补给。

(2)径流条件。山东省深层地下水自西向东或自西南向东北运动，在齐河、济阳、乐陵、滨州及东部滨海地区形成自流区。20世纪70年代后期，德州、滨州、聊城、济宁等重点城镇开始开采深层地下水，经过几十年的开采，深层地下水位大幅下降，形成了以德州、滨州、聊城、菏泽、济宁为中心的区域性深层地下水降落漏斗，以及高唐、临邑、惠民等县城为中心的次级小漏斗，改变了深层地下水的天然流向。目前，深层地下水流向多由各漏斗边缘向漏斗中心区汇流。

(3)排泄条件。深层地下水在天然状态下以径流排泄为主，开采状态下排泄方式以人工开采为主。山东深层地下水的开采主要集中于城镇和工矿集中区，开采区均形成了规模不等的地下水降落漏斗。

(4)人类活动对地下水循环演化的影响。地下水流场变化：20世纪70年代，山东深层地下水开采主要集中在德州市和滨州市，根据1980年的水位观测资料，省内除德州市和滨州市形成小面积的降落漏斗外，其他地区流面平直、流线分布均匀，基本反映了天然流场特征。随着开采量的增加，地下水流场发生了根本变化。目前，山东深层地下水已由天然的自西向东或自西南向东北的统一流场，变为以主要城镇为中心、离散的并各自独立的人工局部流场，水力坡度明显增大。

排泄条件的变化：天然条件下，深层地下水主要是径流排泄和向上部含水层顶托越流排泄。目前人工开采基本成为深层地下水的唯一排泄方式。

3.3 深层地下水水化学特征

深层地下水水化学特征受控于古沉积环境，具有一定的水平分带性。根据舒卡列夫法分类，按阴离子类型分为重碳酸盐型、硫酸氯化物型、氯化物硫酸盐型、重碳酸氯化物型、重碳酸硫酸盐型、重碳酸氯化物硫酸盐型和氯化物重碳酸硫酸盐型；按阳离子类型主要为钠型。

据^{14}C分析资料计算，浅层地下水的年龄均小于4 000年。深层地下水的年龄1.02万~2.10万年，其中惠民县李庄镇深层水年龄最小，为1.02万年，德州市深层水的年龄最大，为2.10万年，其中德州—齐河的同位素剖面，深层地下水的氚值均小于2.2 TU。表明深层地下水多为古封存水。深层水之间存在年龄差异则说明可能接受了新水的补给，成为混合水，但补给源与补给机理有待查明。

3.4 深层地下水动态特征

根据动态监测资料，山东深层地下水的动态特征大致可以分为两类：

开采消耗型：主要分布于德州、滨州、聊城、菏泽、济宁、高唐、夏津、临邑、惠民等城镇

区，深层地下水开采量大、开采集中，已形成规模不同的地下水降落漏斗。受开采影响，年内最高水位一般出现在1~3月份，最低水位出现在11~12月份，夏季地下水开采量大，水位下降幅度相应增大。地下水多年动态呈连续下降趋势，年降幅2m左右。

径流消耗型：分布于漏斗区以外的广大农村地区，区内深层地下水开采量较小或不开采，但受漏斗区深层地下水开采的影响，水头下降。年内地下水头波动变化较小，地下水多年动态呈连续下降的趋势，年降幅0.5～1m。

从1980年至2002年，鲁北平原深层地下水的降落漏斗面积呈逐年增大的趋势，1980年漏斗中心水位为-20m左右，到2002年漏斗中心水位达-70m。

以德州为例，1965年开始将深层地下水作为供水水源，1965年前，深层地下水头埋深2.0m左右。1970年德州漏斗业已形成。到1973年漏斗中心水位埋深已达30.0 m，-10.0 m等水压线圈闭面积为37 km²。到1990年德州城区漏斗中心水位埋深76.23 m，-10 m等水压线范围面积1 707.5 km²。目前，德州城区已建深机井266眼，总开采量3 300万 m³/a。德州城区漏斗中心水位埋深130.16 m，-10 m等水压线范围面积已经达到4 823 km²。

3.5 深层地下水资源

(1)深层地下水系统由第四纪早更新世(Q_1)和新近纪(N_2)明化镇组上部地层组成；地层岩性以黏性土为主，含水层主要岩性为细砂和中砂，砂黏比15%～30%。

(2)水力性质为承压水，地下水运动视为平面二维流，服从达西定律。

(3)补给条件差，循环交替缓慢，开采主要是消耗地下水储存量。

(4)长期超采，水头大幅下降，造成上覆含水层向深层地下水越流补给和黏性土压缩释水。

深层地下水属消耗性资源，开采使水头不断下降。根据水头埋深及降速，参照开采环境约束条件——水位标高为-40m，使用时限为30年。采用水均衡法计算面积59 410 km²范围的资源见表1。

表1　山东省深层地下水可开采资源计算成果表　　　　　　　　（单位：万 m³/a）

计算区	亚区	含水层弹性释水量	弱透水层压密释水量	侧向补给量	越流补给量	小计	合计
鲁西北区	河湖相地下水亚区	12 765	14 733	51	1 788	29 337	33 151
	冲洪积相地下水亚区	3 883		-69		3 814	
鲁西南区	湖西冲积湖积相亚区	3 107	9 362	131	424	13 024	22 117
	湖北山前冲洪积亚区	967	1 150	1 213	441	3 771	
							55 268

4 深层地下水资源开发利用现状与问题

4.1 深层地下水开发利用现状

调查结果表明部分城市深层地下水开采量为：德州3 300万 m³/a，占城市供水总量的19%；滨州428万 m³/a，占城市供水总量的4.5%；济宁12 738.5万 m³/a，占城市供水总量的90%；菏泽1 690万 m³/a，占城市供水总量的25.8%；东营201万 m³/a，占城市供水总量的4.9%。这些地下水主要用于工业生产用水和城市居民生活用水。

4.2 与深层地下水有关的环境地质问题

4.2.1 地面沉降

深层地下水超采造成水头大幅下降是地面沉降的主要诱因。据德州1989～2000年地面沉降监测资料分析，1991年以前，深层地下水头埋深小于74 m时，地面沉降量较小，随着地下水头的持续降低，地面沉降量相应地增加。德州市区地面沉降范围与漏斗分布范围基本一致，特别是城区地面沉降的范围与漏斗范围基本吻合；地面沉降速率的变化趋势是西北部大于东南部，与德州市区深层水开采特征和漏斗区水头特征基本一致(见表2)。

表2 德州市漏斗中心地下水位与地面沉降关系表

时间(年-月)	1989-12	1990-12	1991-12	1992-12	1993-12	1994-12	2000-12
漏斗中心区水位埋深(m)	74.54	76.53	79.26	84.22	86.89	87.86	106.00
水位降差(m)		1.99	2.73	4.96	2.67	0.97	18.14
漏斗中心区地面沉降量(mm)		−31.0	−55.0	−104.0			−387.0
沉降差(mm)		−31.0	−24.0	−49.0			−283.0
水位每下降1.0 m沉降量(mm)		−15.58	−8.79	−9.87			−12.99

产生地面沉降较明显的地区有德州城区、济宁城区、菏泽城区，地面沉降面积相对较大。

地面沉降目前给山东省内带来的危害已较为明显，特别是德州城区表现得尤为突出。主要有以下方面：

(1)城市内涝积水，加大了防洪难度；
(2)路基不均匀下沉，威胁铁路等交通干线安全；
(3)河床下沉，影响输水工程安全；
(4)城市供水、供气管网破坏；
(5)地裂缝频发，危及城乡建筑安全；
(6)浅层水位相对抬升，引起一系列城市环境问题；
(7)河道防洪排泄能力降低；
(8)地面高程资料大范围失效。

4.2.2 串层污染

目前在淄博和东营部分地区已经出现了浅、中、深地下水之间的串层污染现象，主要是水井施工止水质量差造成的。

5 深层地下水开发利用前景

5.1 合理井深、井距和允许降深的确定

合理井深的确定要根据当地深层地下水顶板埋深、主要含水砂层垂向分布情况进行确定，一般钻孔要揭穿主要含水砂层可获得期望的出水量。

合理井间距的确定要根据当地的富水性、互阻抽水开采井水位变化情况进行确定。一般而言，开采深层地下水以分散式开采为宜，可有效避免局部的水位降落漏斗产生。

允许降深根据近年来德州市地面沉降发展变化，当深层地下水头埋深小于74 m时(地面沉降原因部分分析)，地面沉降量较小，随着地下水头的持续降低，地面沉降量相应地增加，由此最终确定允许降深为70 m。

5.2 深层地下水开发利用前景与开发利用对策分析

5.2.1 开发利用前景分析

深层地下水属消耗型水源，形成年龄1万~2万年，补给恢复能力很差，目前的开采量已经处于超采的边缘，但其储存量巨大，在目前没有替代水源的条件下，可作为城镇供水补充水源或应急水源开发利用。但应当严格控制深层地下水资源的开采，特别是新增集中供水水源地应当经过严格论证方可实施。

5.2.2 开发利用对策分析

各个地面沉降区应当逐渐调减开采量，如德州、菏泽、济宁城区；已有水源地以保持现有开采量为开采上限；非应急条件下，不开辟新的集中供水水源地；有条件的地区关停集中供水水源地；积极利用南水北调工程引水，调减深层地下水开采量。

6 深层地下水资源保护

6.1 加强勘查评价与监测

每个水源地可布置 10 个观测点,水位、水温的观测频率可定为每月 3 次,水量和水质的观测可在每年的枯、丰水期各进行一次。对观测数据进行及时整理和统计,分析地下水各要素的变化趋势。

6.2 科学控制开采量

根据相应的水源地勘查评价成果,计算出允许开采量。

6.3 回灌升漏

回灌升漏是把已经疏干的含水层当做一个巨大的地下调蓄水库,本着汛期回灌、冬春季节使用的原则缓解区内水资源供需不平衡的矛盾。

深层地下水进行人工回灌补源可防止大面积的降落漏斗和地面沉降等环境地质问题;其次,深层地下水水质都是高氟、高碘,用黄河水进行回灌,可改善地下水水质,减少地方病危害;第三,地下水库的形成,可给工农业的发展提供水资源保障,对经济和社会发展起到有力的促进作用。

6.4 卫生防护

对废弃井应采取闭井处理,避免二次污染含水层。新打深机井在成井时做好止水工作,以防止串层污染。

河流生态修复——时代赋予水利的重要使命

宫崇楠

(山东水利学会，山东，济南 250013)

1 历史的回顾

中国的水利与中国的历史一样源远流长。大禹治水的传说，可以追溯到 4 000 多年以前，举世闻名的京杭大运河始建于 1 000 多年前。再如郑国渠、都江堰等，我们的祖先在水利发展的历史进程中，曾经创建过诸多的辉煌。这一切，为治国安邦、发展经济、提高人民生活水平，均发挥了巨大的保障和推动作用。但是，与此同时生态环境也受到了不同程度的胁迫，无论是西方经济发达国家，还是发展中国家，大致都经历过或者将要经历以下几个发展阶段：

(1)整治河道，减除水患，以改善人们的基本生存条件；
(2)蓄水引水，发展灌溉，以保障人们的衣食生活需要；
(3)发展城市供水，以满足城市化日益增长的用水需求；
(4)防治水污染，以缓解工业化带来的负面影响；
(5)生态修复，以改善生物和人类的生存环境，实现经济、社会的可持续发展。

现在，越来越多的人开始意识到，为了人类的生存，为了给子孙后代留下一方蓝天碧水，必须转变传统的治水观念，倡导人与自然和谐共存、人与水和谐相处，还河流健康生命。

2 山东省河流生态现状

(1)大多数河流上游都建有蓄水工程，由于水资源供需矛盾突出，被拦蓄的水全部用于工农业和城乡生活，没有考虑下游河道内的生态环境需水。再加上人类活动对下垫面条件的改变，过去由地下水补充所形成的河道基流也不复存在。因此，山东省多数河流在非雨季基本处于干涸状态。

(2)由于城市中的污废水处理、回用的程度低，多数河道都沦为了城市的排污通道，又加上沿途城镇、乡村和农药、化肥等污染，据实测资料分析，约 86%的有水河段，水质都劣于Ⅲ类，其中约 64%的河段劣于Ⅴ类。

(3)流经城市的自然河沟是解决市区排水，保障城市防洪安全的重要基础设施，也是美化城市景观，改善人居环境的宝贵资源。但是，由于对其认识的肤浅和受到各种利益的驱动，填埋、覆盖占用排水河沟的现象比比皆是。如此做法，不但加重了雨洪灾害，而且彻底泯灭了城市的灵气。

(4)尚存的城区河道，大都进行了裁弯取直和断面束窄，几乎都采用混凝土或浆砌石等钢性材料衬砌，横断面也多以直立式挡墙护岸，切断了水与土壤的联系，破坏了河流中生物的生存环境，降低了河流本身的自净能力，绝大多数已经沦为没有生命的死河。

(5)部分河流中、下游的湿地(包括湖泊、洼地等)，已多被围垦、开发，因而不同程度地降低了沿河湖泊、洼地滞蓄洪水、削减洪峰的作用，同时也加重了河流下游的防洪负担。另外，也削弱了其涵养水源、改善生态的重要效能。

3 目前国际流行的河流治理模式

发达国家早在 20 世纪 70 年代，就已经开始关注由于高度工业化和城市化对生态环境造成的负面影响，强调对河流实施物理、化学、生物过程的协调管理，提出了近自然法河道设计理念，并投入巨资进行水污染防治和生态修复。采用生态工程法治理河流环境、恢复水质和生物多样性、维护河岸景观，将河流尽量恢复到接近自然的状态，而且已经取得了显著的效果。目前，国际上流行的多自然型河流治理理念，与传统的以水力学、材料力学、结构力学为基本理论，以输水顺畅为目标的治水模式相比较，两者存在较大差异。前者考虑的基本原则和目标主要有以下几点：

(1)优先修复河流的生态功能，重塑弯曲河道、修复或重建水边湿地，恢复生物多样性，修复河道原有结构，为生物营造多样的、丰富的生存空间。

(2)不提倡单一的渠化断面，允许河岸冲刷、淤积等动态变化。

(3)重建河道与洪泛区的联系，给河流以空间。

(4)提倡"原汁原味"地修复、重建河流富有个性特征和多姿多彩的自然景观等。

4 国内采用新理念综合整治河流的几个范例

(1)上海市水务局提出了"安全、资源、环境"三位一体，协调发展的治水方针，建设自然生态型河道，创建城市水-绿生态体系，建设生态城市的治水目标。主要包括恢复河湖面积总量、优化水网结构、调治水体促进水循环、构筑人工森林湿地、改善水质等，实现城市的可持续发展。规划到 2020 年，可以乘船漫游在市区 40 km 长的环形水系上，亲近水并享受水的欢乐。

(2)北京市近几年以回归自然、恢复生态、生物多样性、以人为本、人水亲近等全新的治水理念，指导水工与景观设计。北京市北环水系综合整治中的转河治理，可称为城市河道整治的成功范例。其设计原则与特色：①历史遗址与现代文化共存；②与城市建筑相协调；③河岸形态回归自然；④恢复生物多样性；⑤以人为本，构筑人与自然的交流平台；⑥人水相亲，感受大自然的美妙与韵味。

(3)绍兴是文坛巨匠鲁迅的故乡，素有"山清水秀之乡，历史文物之帮，名人荟萃之地"盛誉的历史文化名城。近年来，按照"城在水中，水在城中"和"传承古越文脉，展示水乡风情"的建设目标，依据"人水相亲、和谐相处"的设计理念，实现了二环绿水绕古城，打造了具有古今结合特色的江南水城风貌。

(4)太原市近年来对穿越市区原本污染严重的汾河，进行了大规模的综合整治，成为市区的一大亮点，获国家"鲁班奖"和联合国"最适合人类居住奖"。二期工程规划方案，突出了重塑河道的自然形态、修复生态功能和改善环境的理念，现正在全市公开征询意见，准备实施。

5 山东省部分城市开始以新的理念综合整治河流

(1)滨州市结合新区建设，提出了建设生态城市的口号。对现有水系进行了统一规划，实施了大规模的"四环五海"建设项目，将市政、通信、水系、生态、环境等方面有机地结合起来。

(2)淄博市在编制孝妇河综合治理规划时，由水行政主管部门牵头，联合城建、环保、交通、园林、旅游等部门，依照"人水和谐"和"以人为本"的治水理念，从防洪减灾、水资源利用、水污染防治、园林绿化、交通旅游等多个方面，对孝妇河进行了统筹规划。

(3)东营市水利局在广利河城区段的治理规划中，以"传承黄河文化、展现东营文明"为主题，以"以人为本、功能适应、地域风情、生态塑造、经济适用、美中有变、变中求美"的设计原则，沿 4.75 km 长的河道，规划了"湿地公园"、"黄河水利科技文化主题公园"和"智者乐水、黄河之水天上来、大禹治水、棋琴书画、七彩飞虹"等节点景区。

(4)聊城市较早地提出了建设"江北水城"的口号，水系整治与城市景观建设已初见成效。

(5)临沂市利用改建的小埠东橡胶坝构筑大水面，兴建滨河大道，改善市区环境，为市民营造了一个与水亲近的休闲、娱乐场所。

(6)潍坊市近两年先后对穿越市区的张面河、虞河，结合城区改造。通过拆迁清障、河道清淤、护岸截污、调水蓄水、绿化美化等措施，基本达到了建成全国一流滨水景观带。

6 对山东省恢复河流健康生命的几点建议

(1)首先，在全省水资源总体规划以及区域水资源统一调度和优化配置过程中，应该科学合理地安排并优先保障河道内能够维持常年不断流，并且逐步恢复和维持生物多样性，进而达到河道能够经常保持良好的生态环境和景观的基流量(环境流量)。

(2)切实加强各项水污染防治工作。实行最严格的排污总量控制制度，建立重点污染源在线监控网络，加快城市污水处理厂和配套管网系统建设速度，加大运行监控力度，逐步减少农药、化肥使用量，有效控制农业残留物污染。

(3)在河道综合整治规划设计中，将维持河道健康生命作为首要的治理目标。积极采取措施恢复河道和沿岸洼地、湖泊的水力联系，逐步修复河道和岸边的多自然形态，为生物多样性创造必要的生存、繁衍条件，实现健康生命的修复目标。

(4)城市中已被覆盖和填埋的天然河、沟，应尽量予以恢复，将其视为城市的一种宝贵的自然资源。

(5)按照"山东生态省建设规划纲要"的要求，进行全省湿地普查并建立相应信息管理系统，尽快开展各类湿地生态系统的修复和重建工作。

(6)在河道上新建拦蓄水工程或对已建水库、闸坝进行除险加固时，在建筑物的地基渗透稳定计算成果符合安全要求，或者采取简易工程措施即可达到安全要求的情况下，不必将基础处理得"滴水不漏"。这样做可以在保障工程安全的前提下，维持下游河道一定的潜流量，这对于河流的生态修复具有重要价值。

参 考 文 献

[1] 温明霞. 北京市北环水系综合治理工程转河段工程的设计与思考[J]. 北京水务, 2001(3).

[2] 邓卓智. 转河景观设计的原则和方法[J]. 北京水利, 1990(4).

[3] 上海水利学会. 上海城市水利与可持续发展研究[R]. 2004(5).

[4] 卫明. 自然生态型河道建设的理念及其应用[J]. 中国水利. 1999(3).

[5] 刘树坤. 水利建设中的景观与水文化[J]. 水利水电技术, 2003(1).

[6] 刘汉桂. 还清河湖是城市水环境建设的根本目标[J]. 水土保持研究, 2001(4).

[7] 刘晓涛. 城市河流治理规划若干问题的探讨[J]. 水利规划与设计, 2001(3).

[8] 许士国, 等. 现代河道规划设计与治理[M]. 北京：中国水利水电出版社, 2001.

[9] 董哲仁. 试论生态水利工程的设计原则[J]. 水利学报, 2004(1).

[10] 河道整治中心编著. 多自然型河流建设的施工方法及要点[M]. 周怀东,等译. 北京：中国水利水电出版社, 2000.

地下水脆弱性评价方法及其应用

张保祥

(山东省水利科学研究院，山东，济南 250013)

1 地下水脆弱性研究进展

从 20 世纪 80 年代后期开始，在 IHP 和 UNESCO 的推动下，欧美国家陆续开展了地下水脆弱性调查评价与编图。美国环境保护署提出了地下水脆弱性 DRASTIC 评价方法，该方法在美国多个地区获得了成功的应用并积累了丰富的经验，后被加拿大、南非等国采用，用于具有不同水文地质条件地区的地下水固有脆弱性的评价。之后，不少水文地质学家将其用于更大范围水文地质单元的地下水脆弱性评价，并对该系统进行了补充和完善，可以适应各种不同的水文地质条件。后来随着农业的发展，农业污染日趋严重，又提出了考虑农业活动的农药 DRASTIC 标准，该标准用于地下水的特殊脆弱性评价。DRASTIC 地下水脆弱性评价方法于 1991 年被引入欧共体国家，作为欧共体各国地下水脆弱性评价的统一标准。同时，地下水脆弱性评价与编图也成为 20 世纪 80 年代末以来一些国际会议的主题，并出版了许多大比例尺地下水脆弱性图件。目前国外大部分有关地下水脆弱性的研究重点已经转到应用结合地下水质运移模型、统计模型、随机模型等数学手段来评价地下水的脆弱性，多以农药 DRASTIC 标准为基础，结合 GIS 技术来评价地下水对污染尤其是农业污染的脆弱性。

国内于 20 世纪 90 年代初始见报道国外地下水脆弱性研究动态。到 90 年代中期以后，主要是从水文地质本身内部要素的角度来研究地下水的固有脆弱性，且脆弱性研究主要是针对局部区域或城市进行的。从国内应用情况来看，根据各地水文地质具体情况，各种评价模型都有自己的指标体系，地区不同、评价模型不同，所选用的参数也不同。但总体来看，目前主要以应用 DRASTIC 方法或其改进模型为主。

2 地下水脆弱性评价方法

地下水脆弱性评价方法包括经验技术和模型模拟。国内外现有的地下水脆弱性评价方法主要有迭置指数法、过程数学模拟法、统计方法和模糊数学方法。

2.1 DRASTIC 评价方法

地下水脆弱性 DRASTIC 评价方法由美国水井协会(NWWA)和美国环境保护署(USEPA)在 20 世纪 80 年代提出。认为，地下水遭受污染的风险大小取决于污染源和含水层本身所固有的水文地质特性——易污染性。该方法选取对含水层脆弱性影响最大的 7 项水文地质参数评价指标来定量分析地下水的脆弱性，即：D——地下水埋深、R——含水层净补给量、A——含水层岩性、S——土壤类型、T——地形坡度、I——非饱和介质影响、C——含水层水力传导系数。它的应用假设条件如下：污染物由地表进入地下；污染物随降雨入渗到地下水中；污染物随水流动；评价区面积大于 40.5 hm^2。

DRASTIC 指标由 3 部分组成：权重、范围(类别)和评分。①权重：每一个 DRASTIC 评价参数根据其对地下水防污性能的作用大小都被赋予一定的权重，权重值大小为 1~5，最重要的评价参数取 5，最不重要的取 1。各评价参数权重取值的大小要结合具体的评价区域来选定。②范围(类别)：根据对地下水防污性能的作用大小可以将每一评价参数分为不同的范围(数值型指标)和类别(文字描述性指标)。③评分：每个评价参数都可用指标值来量化其评分取值范围为 1~10，分别对应于每一评价参数的变化范围(类别)。地下水脆弱性评价 DRASTIC 指数为以上 7 项指标的加权总和。

根据计算出的 DRASTIC 指数，就能够识别地下水污染敏感区，具有较高脆弱性指数的区域的地下水就易于被污染，反之亦然。DRASTIC 脆弱性指数的最小值为 23，最大值为 226，一般在 50~200 之间。

2.2 DRAMTICH 评价方法

结合我国北方滨海地区的情况和对地下水脆弱性影响因素的具体分析，以指标具有代表性、可靠性、科学性、系统性、层次性、可操作性及指标之间无相关性和包容性作为选取指标的原则，建立了我国北方滨海地区地下水脆弱性 DRAMTICH 评价指标体系。该评价体系包括 8 项指标：地下水位埋深(D)、含水层补给模数(R)、含水层岩性(A)、地下水水质(M)、地形坡度(T)、非饱和带岩性(I)、含水层综合渗透系数(C)及人类活动影响(H)。

与地下水脆弱性评价 DRASTIC 方法相类似，DRAMTICH 评价指标体系也由 3 部分组成，即范围(类别)、评分和权重。地下水脆弱性 DRAMTICH 评价方法的计算结果为研究区以上 8 项指标评分的加权总和。DRAMTICH 地下水脆弱性指数的最小值和最大值分别为 26 和 256。

2.3 其他评价方法

除 DRASTIC 和 DRAMTICH 评价方法外，地下水脆弱性评价方法还有很多。下面对其他常见的地下水脆弱性评价方法作一简单介绍。

(1)SEEPAGE 方法。农业活动区地下水环境污染潜势早期评价(SEEPAGE)模型考虑影响地下水脆弱性的各种水文地质条件和土壤物理性质，包括以下评价指标：地形、地下水埋深、包气带介质、含水层岩性、土壤深度和净化能力。

(2)SINTACS 方法。它使用了与 DRASTIC 模型相同的 7 项指标。该系统能更准确地描述地下水脆弱性系统中各指标在空间上的变化，其结果也更加合理。

(3)GOD 方法。主要用以下三个指标：地下水类型(G)、上覆岩层特性(O)、地下水位埋深(D)。GOD 指数为以上三个评分值的乘积。

(4)AVI 方法。它仅考虑主要含水层上覆各岩层的厚度(d)及其垂向渗透系数(k)的影响。

(5)ISIS 方法。主要考虑以下影响因素：地下水年均净补给量、地形、土壤类型、土壤厚度、包气带岩性、包气带厚度、含水层岩性及含水层厚度。

(6)EPIK 方法。是专门用于岩溶含水层的脆弱性评价方法。考虑到岩溶含水层的特定水文地质条件，包括以下 4 种岩溶区水文地质影响因素：表生岩溶带(E)、上伏表土层(P)、入渗条件(I)及岩溶网络发育情况(K)。

(7)DIVERSITY 方法。主要用于国家或州(省)尺度上的脆弱性制图，对含水层补给形式、补给水的流速大小、补给水的扩散形式等进行综合评价，更适合于对非均质的、各向异性的、存在优势入渗的地质体进行脆弱性评价。

(8)SGD 方法。主要考虑以下参数：地下 1 m 内的土壤含水量(S)、渗透率因子(W)、岩性因子(R)、地下水位以上岩层厚度(T)、上层滞水含水层的附加分值(Q)以及承压含水层的附加分值(HP)。含水层总的保护效果(PT)为以上各项指标的函数。

(9)BTU 方法。用于评价多孔介质含水层的本质脆弱性。该方法能满足欧盟水框架协议要求和充分考虑德国各州水文地质数据的不一致性的影响。

(10)GNDCI-CNR 方法。选择一定数量的水文地质条件按照一定标准来确定地下水的固有脆弱性，该方法具有很大的灵活性，它可随研究区的具体情况作适当调整。

3 应用实例

3.1 DRASTIC 评价方法在泰国清迈盆地的应用

(1)基本概况。清迈盆地主要含水层为新近沉积的非固结含水层，其埋深为 20~40 m、出水量大于 20 m³/h，二级阶地含水层埋深 30~100 m 不等、出水量约 20 m³/h，一级阶地含水层埋深最深达 250 m、出水量 2~10 m³/h。固结含水岩层埋深 20~80 m、出水量小于 2 m³/h。清迈盆地的水文地质特征决定了该地区可以应用 DRASTIC 方法来进行地下水脆弱性评价。

(2)地下水脆弱性分区 根据上述 DRASTIC 评价方法，利用 ArcView GIS 作为计算工具，对清迈盆地中清迈府地下水的脆弱性进行了分区，共分为高、中、低三个档次。

清迈府的地下水高脆弱性分区主要分布在清迈盆地腹地，占总面积的 17.65%，一般都是在地下水位埋深较浅、岩层渗透性较强、地下水补给量较大的地区。低脆弱性分区主要分布在清迈盆地的西部地势较高且地形坡度较大的地方，占总面积的 52.50%。中脆弱性分区占总面积的 29.85%，主要分布在宾河支流区域及部分山间盆地，本区地下水位相对较深，是主要的农业耕作区。

在浅层含水层不发育的地区，其覆盖层的连续性较好，主要含水层的埋深一般在 30 m 以下，地下水一般不易受到污染，其脆弱性较低。因此，可以建设垃圾处理场、污水处理厂等有可能对地下水造成影响的项目。

3.2 DRAMTICH 评价方法在黄水河流域中下游地区的应用

3.2.1 基本概况

龙口市城区(黄城)位于黄水河流域内西北部。本区最具供水和保护意义的地下水为位于王屋水库下游至黄水河地下水库坝址之间的范围。评价区范围在南部和东部大致以构造剥蚀低山丘陵区与剥蚀堆积山前台地交接地带的 80 m 等高线为界，在西部和北部是以黄水河流域界线为界；在东北部流域界限至龙口市与蓬莱市县界处，此处地表水系直流入海，但地下水在被黄水河地下坝工程拦截后直接汇入黄水河地下水库，因此将该二区也划为评价范围。在评价区域内，仍有局部高程在 80 m 以上孤立的低山丘陵存在，主要分布在评价区西南部及东部，面积较小。评价区内除了分散取水外，已经建成了两个集中供水水源地。评价区域总面积为 319.97 km²，占整个黄水河流域总面积的 30.9%。

3.2.2 地下水脆弱性分区

根据前述的 DRAMTICH 评价方法，利用国产地理信息系统软件 Map GIS 对研究区进行地下水脆弱性评价。首先利用 Map GIS 的地图编辑功能，对研究区地下水脆弱性评价 DRAMTICH 方法的各个指标分区分别进行数字化，使其可以进行编辑。对各个指标分区图分别进行配准后，把每个指标的范围(分类)和评分值分别赋予其相应的属性参数中去，这样利用"地图编辑"模块就可以清楚地对各指标分区图进行显示和编辑。再利用"空间分析"模块对研究区地下水脆弱性评价 DRAMTICH 评价体系 8 项指标的分区图分别进行空间合并操作形成新的分区，其总评分值在 32~227 之间。

根据计算出的 DRAMTICH 总评分值，具有较高脆弱性分值的区域地下水易于被污染，按照研究区地下水脆弱性评价 DRAMTICH 方法总评分值对新形成的分区进行再分类，最终完成黄水河流域中下游地下水脆弱性研究分区图，用于指导城市规划和工农业生产布局，为决策和管理部门制定地下水保护策略提供依据。根据龙口市和黄水河流域的实际情况，将地下水脆弱性分为 5 个等级：①极高脆弱性，总评分值为 227~189；②高脆弱性，总评分值为 188~158；③中等脆弱性，总评分值为 157~111；④低脆弱性，总评分值为 110~72；⑤很低脆弱性，总评分值为 71~32。

4 结语

近年来，国内外在地下水脆弱性评价与编图方面做了大量工作，取得了许多理论和实践成果。地下水脆弱性评价与编图已成为现代水文地质学的新分支，同时也是保护地下水资源和实现其可持续开发利用的有效技术手段。但目前在地下水污染脆弱性评价与编图方面还存在一些认识问题或薄弱环节，有待于进一步完善和提高。

参 考 文 献

[1] 孙才志,潘俊.地下水脆弱性的概念、评价方法与研究前景[J]. 水科学进展,1999, 10(4).

[2] 姜桂华.地下水脆弱性研究进展[J]. 世界地质,2002, 21(1).

[3] 杨庆, 栾茂田. 地下水易污性评价方法—DRASTIC 指标体系[J]. 水文地质工程地质, 1999 (2).

[4] 钟佐燊. 地下水防污性能评价方法探讨[J]. 地学前缘,2005, 12(增).

[5] 徐月珍. 法国的地下水易污性编图[J]. 水文地质工程地质,1990(1).

[6] Michael J. Focazio,Thomas E. Reilly,Michael G. Rupert,et al. Assessing Ground-Water Vulnerability to Contamination: Providing Scientifically Defensible Information for Decision Makers[J]. U.S. Geological SurveyCircular 1224. 2000(4).

[7] Michael G. Rupert. Improvements to the DRASTIC Ground-Water Vulnerability Mapping Method[J]. National Water-Quality Assessment Program–Nawqa,USGS Fact Sheet FS-066-99,1999(6).

[8] Jaroslav Vrba, Alexander Zaporozec. Eds. International Association of Hydrogeologists. Guidebook on Mapping Groundwater Vulnerability[J]. International contributions to hydrogeology,Vol.16,Hannover: Heise, 1994(2).

Assessing the State of Water Resources and Sustainable Water Management Strategies in Taicang City, Jiangsu Province

Lu Zhibo　Wang Juan　Deng Dehan　Liu Ning　Yang Jian
(School of environment science and engineering, Tongji University, Shanghai, China)

Taicang City, is on the beginning of the turning point towards the basic modernization, as well as the key to achieving sustainable development in the early stage of new century. Therefore, Taicang City CPC Committee and Taicang government, created the"Eco-city economic cycle city"strategic goal, in order to reasonable arrangements and plans for the relationship between social, economic and environmental resources in urban development, ensure the safety of the city's ecological system, sustainable economic growth and the healthy of social development. Water is the basis of Taicang City, but water quality is poor, more water resources for the transit of water, upstream runoff. Large amount of wide range non-point source water pollution, those lead the urban ecological deterioration. "More water management" the traditional water management situation has increasingly shown its drawbacks.

As a new, integrated water management concept, sustainable water management systems have not been built up international and national. On the basis of in-depth analysis of this situation and existing problems in the water in Taicang City, this article try to develop a suitable strategy for sustainable water management system.

1 Water Resources Assessment

Southeastern location in Taicang in Jiangsu Province. Located N 120° 58′~121° 31′, E 31° 22′ and 31° 44′. East brink of the Yangtze River, across from Shanghai Chongming County, on the south of Baoshan District, Jiading District, west Kunshan City, north Changshu City.

1.1 Interannual variability of precipitation

Wet and dry precipitation is shown in Figure 1.

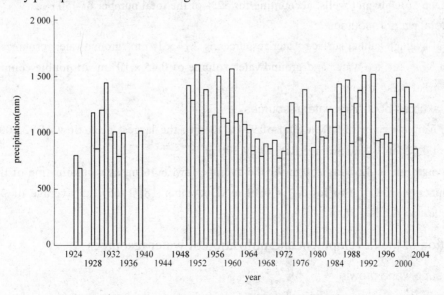

Figure 1　Precipitation in Taicang City

1.2 Seasonal changes in precipitation

Because of the monsoon climate, the annual precipitation and seasonal changes are large, generally speaking, more rainfall in spring and summer precipitation, less Rainfall in the winter; during the past 20 years, from 1984 to 2003 the biggest monthly precipitation happened in June 1999. Reached 600.2 mm, the smallest monthly precipitation happened in December 1987, no precipitation; maximum consecutive days without precipitation in November 29, 1987 ~ January 1, 1988, 34 consecutive days without precipitation.

1.3 River runoff

Taicang is the Yangtze River Delta alluvial plain, the city's, total area is 822.9 km^2 water area is 271.0 km^2, the waters of the Yangtze River area is 173.9 km^2. Land area is 537.0 km^2. Taicang City located in the southern Yangtze River, the Tai Lake downstream, it belongs to the Yangtze River delta new plain and lower terrain, east slightly higher Western low. There are 4,110 rivers and 4,213.4 km long.

The local average runoff is 176×10^6 m^3, the average displacement is 1.414×10^9 m^3. The average water content is 713×10^6 m^3. River average water levels is 2.72 m, highest level is 4.07 m (August 4, 1960), the lowest level is 2.12 m (February 10, 1956).

1.4 Groundwater resources

Taicang group is a loose salts pore water delta; underground water-bearing is weak. Groundwater is divided into shallow groundwater (diving), and deep groundwater (water pressure) groundwater, According to hydrological and geological conditions of Taicang City, Part II of confined water main mining of groundwater levels, I and III confined water level to support mining.

According to the survey data, groundwater is not rich in Taicang, poor water quality. Single Well 25~60 m^3/h of water. Taicang City, groundwater resources reserves of 208×10^6 m^3, the average annual groundwater resources is 138×10^6 m^3. 2000 actual extraction of groundwater is 0.35×10^6 m^3, of which 0.19×10^6 m^3 of industry, 0.16×10^6 m^3 of life.

After Taicang was included in the Prohibited Area in 2000, the amount of groundwater exploitation has been effectively controlled. In 2001, the total volume of 0.28×10^6 m^3, in 2002 is 0.19×10^6 m^3. Compare to 0.35×10^6 m^3 in 2000, a total of two years to abate 0.24×10^6 m^3.2001, 2002 Taicang City closed 243 deep groundwater wells, accounting for 52% of the total number of 450 eyes.

1.5 Local total water resources

Historical average annual surface water resources is 2.14×10^6 m^3, groundwater resources is 1.38×10^6 m^3. Removing surface water and groundwater volume of 0.45×10^6 m^3 of double counting, a total of 307×10^6 m^3 of water resources.

1.6 The total cross-boundary water resources

Taicang Yangtze river runoff, a total of 924×10^9 m^3, the largest peak flow of 92,600 m^3/s. The lowest flow of 4,620 m^3/s, the average flow of 29,300 m^3/s.

The average tide is 3.62 m, 1.50 m for the average low. 6.46 m, as a culmination of the calendar year (September 1, 1981). The lowest is 0.49 m (December 18, 1965), an average of 2.12 m tide distance, biggest splash is 4.38 m.

2 Evaluation of the development and utilization of water resources

2.1 Water supply and growth

There is one water company in Taicang City, which is the water treatment limited liability company, There is one water plant under it (the second water plant) and 158 transfer water plant, the total supply capacity of 220,000 m³/d. Water is transported to all the areas in Taicang except Shuangfeng, Xingchao and Zhitang. The water intake installed in the Yangtze River port where has clean, enough water.

2.2 water consumption and growth

According to statistics which don't include ecological water system traditionally, water consumption includes industrial production water, agricultural water and daily life water of the residents. Agriculture (especially farming) water fluctuates inversely with changes in precipitation, and industrial water consumption. Urban water and rural life surface water upward trend, because of Taicang City industrial development, the rapid growth of urban population, the rise in the living standards of farmers.

Table 1 Taicang City, water use efficiency and water structure in 2000 to 2003

Indicator	Unit	2000	2001	2002	2003
GDP	10,000 yuan	1,563,134	1,579,805	1,800,693	2,100,000
Total water	10,000 m³	29,810.93	27,895.19	30,870.39	27,561.96
Agriculture	10,000 m³	15,770	13,151	12,907	11,641
Industry	10,000 m³	10,653.51	11,340.13	14,540.13	12,464.7
Life	10,000 m³	3,387.42	3,404.06	3,423.26	3,456.26
Water per GDP	m³/10,000 yuan	190.7	176.6	171.4	131.2

2.3 The main problem in the development and utilization of water resources

(1) Water shortage situation intensified.

(2) Phenomenon that water for cities occupation agricultural and the ecological environment, water uses has become increasingly prominent.

(3) Wasting water and water scarcity coexist, and there is a great potential for water conservation purposes.

(4) Worsening water pollution, surface water and groundwater quality has declined.

(5) To raise the level of water resources management.

2.4 Surface Water Environment Assessment

2.4.1 Potable water evolution

The results: Since 1996, drinking water sources meet Ⅲ class water standards (sub-index greater than 60). Groundwater quality is better than surface water and the Yangtze River water is superior to other surface waters. Taking into account the excessive extraction of underground water, taking the Yangtze River water as water resource is more reasonable and reliable. From a management point of view, water quality monitoring project and the frequency has increased. From the original 11~27 in 2003, a more comprehensive reflection of the local water quality. From 1997 to 1999 the groundwater monitoring data sources can be seen, there is a degree of local mercury pollution. Reasons to be further confirmed. Since 2001, there has been a decline in the water quality of the Yangtze River water. 2003 did not meet the requirements for the project followed by ammonia (Ⅲ), phenol (Ⅲ). Should cause a certain degree of attention.

2.4.2 Water evolution of the major rivers

(1) Since 1996, the river water quality is generally good (sub-index greater than 60). Before 2001

due to the lack of ammonia nitrogen, total phosphorus, total nitrogen, chemical oxygen demand and other indicators in monitoring data, Evaluation results have certain limitations, but the basic factor for the sub-standard is oil, phenol, oxygen organics. Since 2001 it gradually increase ammonia, chemical oxygen demand, total nitrogen, total phosphorus and other conventional monitoring data, sub-standard factor remains oil, phenol and oxygen organics. According to the monitoring data from 2001 to 2003, it can determine the region's main river pollution factor for ammonia, total phosphorus, chemical oxygen demand, oil and phenol, there should be a focus on identifying reasons.

(2) Sub-index. Taicang City's comprehensive water quality analysis on the evolution trend of positive correlation with the number, y sub-index for river water quality, x years, there is $4.741,2\ln y = x + 65.927$. The correlation coefficient $R^2 = 0.888,4$, and the higher the correlation.

2.5 Groundwater environment situation appraisal

From 1997 to 1999, the water quality of groundwater source. As said above, the Taicang City from 1997 to 1999 the tap water of water source is the groundwater, the water quality fractional exponent in 1997 is 86.8, in 1998 is 100, in 1999 is 88.4, carried on the synthesis appraisal according to the GB 3838—2002 standard, the groundwater water quality overall achieved II kind of water body standard, has not reached the sign project for the total mercury (IV kind).

3 Main question and restriction factor

(1) Average per capita local water resources insufficient.

(2) Water supply varies greatly from year to year.

(3) Transits the water volume to be rich, but it is influenced by salty tide.

(4) Groundwater excessive mining, the water level drops.

(5) The water pollution further causes the water resources crisis.

(6) The influence in the warming climate.

(7) The population increase and the economy grow dual pressure.

(8) Unreasonable economic structure and extensive development patterns.

(9) Lower rate of wastewater treatment and reuse.

(10) Non-point source pollution is very serious, but has not taken effective measures to.

(11) Meteoric impact on the quality of weather pollution.

(12) Management mechanism is imperfect and insufficient capital investment.

4 Conclusions and Outlook

Water is an indispensable element to all human beings and all the other organisms, as well as irreplaceable extremely valuable natural resources for industrial and agricultural production, economic growth and social progress. It constitutes the basic condition for sustainable human development with air, land and energy.

With population growth and rapid economic development, and water pollution has been increasing, the availability of water are becoming scarce. Many areas of the global water crisis. Water shortages have constrained economic development and people's survival food production, but are directly harmful to people's health. Competition for water resources, water pollution control, and even some parts of the Earth international conflicts. Water is an objective existence, for the overall system. The provision does not exist in human society, including the various administrative boundary lines. But to truly reflect the impact load imposed by the human society. To further enhance the level of China's water management, and actively promote the development of China's water management changes from the past and present "static" to "dynamic management"; from "experiential management" to the "scientific management"; from the "scattered fractions management" to "integrated management" changes to support sustainable water management for sustainable socio-economic development, and establish a watershed or regional natural (rain, surface water, groundwater and soil, etc.) and the sustainable use of water and drainage water management systems, it is very necessary and urgent. Taicang City, sustainable water management system is "more water management" reform "a dragon Water", led by the Chief Executive, the functions of various government departments. Sustainable Water Management Coordination Committee, regional flood control, drainage, drought, water and water supply, drainage, water, sewage treatment and recycling, water use, groundwater recharge. Rivers, soil and water conservation, protection of water resources pose a whole affairs branch, in accordance with the principles of sustainability. Under unified management and supervision.

With population growth and rapid economic development, and water pollution has been increasing, the availability of water are becoming scarce. Many areas happen water crisis in the world. Water shortages have constrained economic development and people's survival food production, and are directly harmful to people's health. To compete for water resources, control water pollution, even some parts of the Earth break out international conflicts.

Water is an objective existence, for the overall system. The provision does not exist in human society, including the various administrative boundary lines. But it truly reflects the impact load imposed by the human society. To further enhance the level of China's water management, and actively promote the development of China's water management changes, from the past "static management" to present "dynamic management"; from "experiential management" to the "scientific management"; from the "scattered fractions management" to "integrated management", to support sustainable water management for sustainable socio-economic development, and establish a watershed or regional Natural (rain surface water, groundwater and soil, etc.) and the sustainable use of water and drainage water management systems, it is very necessary and urgent.

Reform of sustainable water management system in Taicang City means, change "more water management" to "a dragon manage Water." That is to say, led by the Chief Executive, organize various government departments to Sustainable Water Management Coordination Committee, make regional flood control, drainage, drought, water resource and water supply, drainage, water conservation, sewage treatment and recycling, rainwater usage, groundwater recharge, rivers, soil and water conservation, protection of water resources to a whole affairs branch, in accordance with the principles of sustainability, practice management and supervision.

Control of the Water Resources Risk

Tian jinghong
(Beijing Technology & Science University, Beijing, China)

1 Water resources safety

Water resources is not only the necessary ware for survival of human being, but also the generation for environment protection and economic development. The change of amount of water resources will cause the relevant water risks, which can be classfied three categories, they are the flood and drought risk, water shortage and environment pollution risk as well as economic risk.

1.1 Flood and drought risk

Global changing, reducing of the plant cover rate, destroying tropic rain forest as well as water and soil errsion and so on, all of these will cause flood and drought increase in the world, this kind of tendency is going to rise. e.g, in the past decades, the flood and drought risk in China is creasing in accordance with the satistic.

The flood and drought has a seasonal charactertistic. Generally the loss of the flood and drought is larger than other disaster, covering a large area at same time. For instance, the flood event, happened from July to August in 1954 in Yangtze River Basin, has destroyed agricultural fields, buildings, manufactories in province Hunan, Hubei, Anhui, Jiangxi and Jiangsu, more than 33,169 persons were dead with a total economic loss of 10×10^9 Yuan. In the same year of 1954, the flood from June to August in Hehe River Basin has made a economic loss of 1.64×10^9 Yuan; the flood event of Hui River basin in 1954 caused 3,014 persons dead. Alltogether, more than 38,000 persons were dead in this three large floods in 1954. In the national safety report in 2002, the recorded economic loss regardingflood events in 2002 was 88.84×10^9 Yuan with a death of 1,171, corresponding to 1 % of total death number of different risk in a year.

1.2 Water shortage and environment pollution risk

The influence of the global warming will cause the subsidise of glacier and the reduction of the permanent snow as well the discharge change in the river basin. The forcast shows that in the near furture, due to the global warming, the river discharge in north part of Hui River Basin, Hehe River Basin and in west of China area will reduce dramatically. The shortage risk in Hehe River Basin, Yellow River Basin and Hui River Basin will increase, which will cause the drease of the purification capacity of river clearly.

The environment pollution will make this situation worse. According to the *Liaowang News Week Report* from 28th of November in 2005, water pollution problem exists in whole country; Water quality in most of surface water quality monitoring station cannot reach Chinese class Ⅲ; 90% of river, which flow through cities, were polluted seriously. Eutrophcation has been happening in most of the lakes in China; there are more than 300×10^6 farmers drinking unsafe water, leading to 80% disease relating to drinking water quality. *The Chinese geological environment* reported that there exist different diseases regarding drinking water quality in 31 provinces, for instance, cancer village in Hui River Basin, more than 120 cancers cases in a few villages in Beichen zone of Tianjin City, malignancy zone coastal area of Shandong Province, groundwater pollution due to solid waste in Beijing, the water pollution due to the explosion of benzene manufactory in Jilin Province in 2005 formed 130 km river pollution zone, which has leaded to stop water supply for 4 days in Harbin City. All of these increase water resources safety risk.

1.3 Economic risk regarding water resources

It is stated in the declaration of water and sustainable development worldwide in Dublin in 1992, "water has own value, water pollution and environment is due to misunderstanding of its economic value"; the agenda of UNCED 21 century in 1992 stated that water resources should be recognized a limited resource with the important sco-economic value.

Compared with developed counties, the technology of water saving in developing counties is relative poor. In 1995, the water using for 10,000 Yuan GDP in Japan, Germany, South Korea, France are 12 m^3, 60 m^3, 43 m^3 and 77 m^3 respectively. In 2002, the water demand for 10,000 Yuan GDP in China is 241 m^3, 5 ~ 10 times higher compared with developed counties. The efficiency of irrigation water is only 40% ~ 50% of the rate in developed counties, the percentage of water recycling rate has long distance to get 80% recycling rate in developed counties.

Although most of cities in China has been implementing water price system for water taking, but according to statistic of "the economic implementation report first 8 months in 2005", the water supply industries are still running under operation cost. Compared with developed counties, the water price still does not reflect the water resources as a limited and valuable source. In " world water report in 2002", the water price in Germany, Denmark, Belgium, Holland, France, England, Italy, Finland and USA are 1.91 \$/m^3、1.64 \$/m^3、1.54 \$/m^3、1.25 \$/m^3、1.23 \$/m^3、1.18 \$/m^3、0.76 \$/m^3、0.69 \$/m^3、0.50 \$/m^3 respectively. From increasing tendency of water shortage and water demand, water price is still lower. Clearly current water resources will not be able to satisfy the requirement in the future. If we do not use our water resources reasonably, the limited water resources will increase the industrial cost, which will lead to reduce of competition capacity of our industries, furthermore, water resources will bring risk for national economic development.

2 The problem of water safety

The flood and drought risk, water shortage and environment pollution risk is increasing with global change. Experts predicts that the problem of water safety is the most important problem in China following the energy problem, which will influence the national target achivment of sco-ecomonic in 21 century.

The safety means no dangers, no disaster happening; water resources points the fresh water resources in the world; the water safety means that people try their best to reduce flood and drought risk, water pollution risk as well as saving water resources through scientific forecast, sysmatic planning, reasonable use, it also means that the water system will ensure the life of human being through warning analyse, countermeasure research under different dangers water risks.

2.1 Water risk

In general, the water risk can be divided into four categories: quantity risk, quality risk, equipment and facilities risk and human activities risk.

2.1.1 Water quantity risk

The uneven quantity change of water in special interval will cause the risk of water safety. The flood risk will happen in heavy rainfall event in a short interval, on contrst, no rainfall in a relative long time period is big disaster of drought. For instance, from 2nd to 5th of september in 2004, the total rainfall was more than 100 mm in Dazhou area in Sichuan Province, there were 67 persons death. Due to global change, the rainfall has been decreasing since 2003 in Heihe River Basin, the serious drought has been leading to the groundwater level droping down 3 m/s in Beijing.

Because of the complication of the climate forecast, water quantity is difficult to predict in accurate for long time. Nomrally people often know its harm after happening. In our country, the heavy rainfall used to happen in summer period, meanwhile, the drought happens in other area. It is a special bad situation during the most important growth period of crops without rainfall, which will lead to the reduce of agricultural products. The change of water quantity could cause other disaster too.

2.1.2 Water quality risk

The water quality risk usually is caused due to huamn activities so that water quality can not reach the national standard, which leads to environment pollution, water shrotage. The statistic shows that 80% of disease is related on the drinking water quality. The bad water quality will destroy water system and bio-diversity.

Clearly, the water quality risk is due to neglection of environment protection of human being. Some manufactories do not want to invest money for pollution control; farmers use more chemic fertilizer in the field, the urbanization leads to urban river pollution; the landfill of solid waste causes the groundwater pollution ect..

The water quality risk is a process of accumulation. The nutrients accumulation of N, P leads to the eutrophication in lake, but the risk of water quality is controllable. With new technologies, the treated wastewater can be reused for green land irrigation, car washing as well as other purposes so that it can reduce the pressure of regional water shortage.

2.1.3 Unstable status risk of equipment and facilities

The unstable status risk of equipments means that the equipment or facility in reservoir, dam, well equipment, water plant, wastewater treatment plant are with unsatble status to cause risk happen and water resources loss. One investigation shows that 10% safety disaster was caused due to unsable status of equipment and facilities. On 8th of August 1975, the Banqiao Reservoir is collapsed due to long running under dangious status. The flood formed 150 km length from east to west, 70 km width from south to north. Till now, there is no information about how many residences have been dead in the disaster, there exist 30,000 dangious reservoirs in whole countey.

The unstable status of matter is growing with implementation gradually, furthermore, causing disaster. The kind of the unstable status of matter is related on the management level and design level; for instance, the water loss in the pipeline of water supply is more than 50% to 60% due to equipment quality and construction technologies.

The unstable status of matter includes lacking of management experiences and lower investment for implementation, as well as unreasonable planning, unsuitable equipment.

2.1.4 Human behavior risk

Human behavior will cause water risk. According to the risk management theory, that intersection of unstable status of the matter and human unsafe behavior is the risk happening point. The unstable equipment and human operation mistake caused the explosion of benzene in Jilin Province in 2005. This kind of unsafe behavior includes the wastewater to be discharged into river without treatment, destorying equipment with purpose, war and so on. Human unsafe behavior connects with the unstable status of matter. Human activities are the main reason for environment pollution.

2.2 Transfer of water risk

The problem of water resources safety is caused by water quantity risk, water quality risk, unstable status of matter and human unsafe behavior or signal risk.

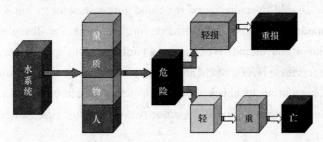

Figure 1   Map of water risk translation

The uneven distribution of water quantity leads to the water risk, the polluted water quality makes this situation going worse. Due to unreasonable planning of water project and lacking of facilities, rainfall in summer period is discharged without reuse, human unsafe behavior causes the unstable status of matter so that it will increase water risk.

The four water risks could cause water disaster, leading to equipments and facilities destory in lighter level and serious level, leading to person light wound, serious wound and death. The degree of water disaster can be described in the following formula.

The degree of disaster={Quantity} ∪ {quality} ∪ {unstable status of matter} ∪ {human unsafe behavior}

3 Control water risk

According to the time order, the water risk can be divided three phases: prophase of water risk, metaphase of water risk and end phase of water risk.

The control of prophase of water risk should include follows.

Estimation and analyse of the possible existing water resources risk, there is:

(1)Control and warning of possible water risk.

(2)Separation and transfer of the existing water risk.

The controls of metaphase of water risks are:

(1)Forecast and warning of the water risk process.

(2)Emergency aid of water risk.

(3)Restoration of water risk disaster.

The controls of the end-phase of water risk include:

(1)Restoration of disaster.

(2)Rebuild of the disaster loss.

(3)Re-estimation and reinvestigation of water risk.

3.1 Water risk control in prophase

The control of water risk in prophase is the most important thing, through analyse and forecast in prophase, the most of water risk at the beginning phase can be controlled.

Through construction in th past, the network of metrological, hydrological and geological monitoring station has been built up draftly, which has been collected a lot of metrological and hydrological data as foundation for forecast and warning of water risk. But the forecast model should cosider the relationship between global climate change model and regional geo-hydrological model. Engineer should use advanced technologies such as radar, satellite, GIS, GPS to build up automatic monitoring system in order to enhance the forecast level.

Not only Engineer should do more work for flood forecast model, but also for drought model, especial hydrological model in aird and semi-aird region. In China, the drought model in Heihe River Basin, Yellow River Basin and inland river basin is badly needed, meanwhile, the forecast of water quality in important reservoirs, rivers and water resources protection zones should not be neglected. People should made a regulation to check the human being unsafe behavior as well as the unstable status of equipments and facilities regularly in order ro avoid water risk caused by Human being and unstable status of matter.

- Control and warning of possible water risk.
- Separation and transfer of the existing water risk.

3.2 Water risk control in metaphase

The water risk control in metaphase includes short time forecast, arrangement aid activities, and effective guide during the water disaster.

The early forecast for risk is the most effective method to reduce loss of the water disaster. For instance, the director economic loss has been reduced 800×10^6 Yuan due to accurate flood forecast in 2002. On 19th of June, Bureau of Hydrology and Water Resources Hunan gave a forecast before 9 hours that water level would rise to 104 m, exceeding 1 m of the warning water level. One suggestion has been gaven for reservoir adjustment. The local government took an activities to adjust this flood successfully. The eocnomic loss was reduced 50×10^6 Yuan. After explosion of benzene manufactory in Jilin Province, 130 km pollution zone in the river moved to the downstream, the concentration of pollutants had been monitored and forecasted timly, the accurate short time forecast provided chance for preparation of cities in the downstream of Songhua River, the economic loss has been reduced in the lower level.

The short time forecast should include drought forecast. Engineer should try best to aviod mistake of forecast, which can cause bad influence in socitey, for instance, the meteorological station gave a forecast during the typhoon activities in Beijng in 2005 that the largest rainfall could happen in Beijing. Beijing government had made a good preparation early, a lot of automic water level equipments had been installed in the main street, less than 1×10^6 inhabitants were ready to move. In fact, there was only normal small rainfall event in Beijing during this typhoon activities.

- Emergency aid during water risk.
- Restoration of water risk disaster.

3.3 Water risk control after disaster

The water risk control in after disaster should include disaster restoration (see Figure 2), reconstruction and re-analyse.

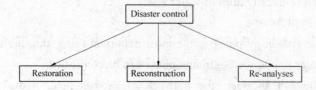

Figure 2　Restoration of water disaster

The restoration of water disaster includes the restoration of daily life and work of human being, mental restoration, restoration of disaster loss. The government should do good job at all of work for restoration as well as statistic of disaster. After disaster, the mental destroy of inhabitant often has been happened due to the dead of relative and economic loss. The mental restoration could encourage inhabitants to increase confident to overcome the difficulties.

The sanitation work and control of disease after disaster should not been neglected. It can be done through the different way to teach the inhabitant to increase the self-capacity for survival, encouraging inhabitant to attend the insurance to get repayment through insurance process. At same time, we should try to get aid from international organization too.

- ●Reconstruction of water disaster.
- ●Re-estimation and re-analyses of water disaster.

4 Outlook

The Global warming and global economic development will cause more disasters, which has related to water safety. Clearly water safety will become other bigger problem behind the energy problem in China. Based on summarizing research experiences, using advanced technologies, it can be expected that the loss and death of water disaster will reduce. Water safety will become more and more important issue for state security and economic development.

References

[1] United Nation, international water development report, 2000: 3-5.
[2] 姜文来. 中国 21 世纪水资源安全对策研究[J]. 工业用水与废水, 2002 (3): 8-12.
[3] 阮本清, 梁瑞驹, 王浩, 等. 流域水资源管理[M]. 北京: 科学出版社, 2001.
[4] 金龙哲, 宋存义. 安全科学技术[M]. 北京: 化学工业出版社, 2004.
[5] 隋鹏程, 陈宝智. 安全原理[M]. 北京: 化学工业出版社, 2005.
[6] 陈绍金. 对水安全文化建设的思考[J]. 中国水利, 2004(13).
[7] 夏军, 黄国和, 庞进武, 等. 可持续水资源管理——理论·方法·应用[M]. 北京: 化学工业出版社, 2005.
[8] 郑通汉. 论水资源安全与水资源安全预警[J]. 中国水利, 2003(6).

The Application of Geographic Information System in Urban Water-supplying and Water-draining Industry

Zeng Wen   Wang Xiguang

(Wuhan Zongdy Cyber Group Co.Ltd, Wuhan, Hubei, 430074)

1 Introduction

Water-supplying and water-draining pipeline is a pipeline network with complicated structure and gigantic scope, which distributes underneath the city's ground freely and is the most important aspect a city depends on. And the safe and efficient management to it is very important to city's safety and development. Each city in our country has a lot of pictures for water-supplying and water-draining's design, construction and completion as well as various pamphlet, and in a long term, the water-supplying and water-draining industries for each city continued using old manual way to manage these data. But this manually manage scheme is now so hard to satisfy real requirement due to the fast development of city's construction, so adopting new technologies and methods to manage water-supplying and water-draining pipeline is one of the most important urgent affairs. This paper introduces the basic theories and importance of GIS, as well as its application in water-supplying and water-draining industry management, and it discusses the future development and research combined with those common issues in water-supplying and water-draining management.

2 The basic theories and importance of GIS

GIS is a kind of system used to collect, store, and process, analyze and display spatial Information. It efficiently manages graphic information and trait data stored in database, including the following aspects.

Data input: used to collect space data which comes from existing maps, remote-sensor and some other aspects.

Data storage and query: used to store space data in various formats, and ensure the database to be updated and corrected swiftly and exactly.

Data processing and analysis: it executes data transferring according certain rules defined by users, or gives variables and constraints of all kinds of time-space optimized models or stimulant models.

Data output: it displays primitive database and processed data, and displays the output of space model in the form of tabulations and maps.

There are several kinds of schemes for collecting GIS data, including manual digital, commercial digital, satellite data and automatic scan. The data's storage in GIS is usually accomplished by using vector structure or griddling data structure. Griddling data structure is also called grid form. One grid or griddling system could be described as a two-dimensional reference frame with rows and lines, and vector system could be described as a space with dots, lines and sides. In GIS, griddling structure is better than vector data structure, that's because all kinds of data information, including various traits data, image data, scanned data, digital terrain model, can be recurred by griddling. Of course, no matter which form you use, it's up to practical needs.

Pipeline information and terrain, environment information that related to pipeline is ultimately geographic information, which has various traits like regional distribution, huge data magnitude, various information carriers. As we always need to execute query and analysis associated with space and topology, we must use geographic information system (GIS) technologies. And we can manage spatial position, attribute trait and time-region trait associated with geographic information unitively by GIS technologies, in order to analyze and generate new geographic information efficiently.

Management information system has to manage spatial data of big magnitude efficiently and display it swiftly, generate various pictures with high quality. it has to generate and manage topology relationship automatically. it has to be able to implement statistic query, spatial analysis (polygon overlapped-position, cache analysis, network analysis) associated with space, as well as 3-dimensional model analysis. it has to provide various spatial data input and output methods. You can't imagine how hard it is to realize those features above without the support of GIS technologies. As a result, water-supplying and water-draining information system must be a GIS application system.

To build water-supplying and water-draining information system on the basis of GIS, you need to choose the GIS platform software it runs. Water-supplying and water-draining information system a application system built up after a second-development on GIS platform. GIS platform software is the underlying support and running environment for water-supplying and water-draining information system, it has great influence for running environment, so choosing a proper GIS platform type is very important. The influential domestic GIS platform products in current market are MAPGIS, Gemstar and so on, abroad products are ARC/INFO, Micro Station.

From the current application condition, the writer thinks that domestic GIS platform has many advantages, such as its interface is slim and beautiful, it fits our people's habits, it uses windows operating system as the operating environment and its price is lower. Some of its features are very close or better than those of abroad software, so it can meet the request of water-supplying and water-draining pipeline system and can be your preference.

3 The main application of geographic information system

According the management requirement for water-supplying and water-draining industry, GIS is usually used in the following managing aspects.

3.1 Daily management for water-supplying and water-draining pipeline

Nowadays, the application of GIS in domestic water-supplying and water-draining industry is mainly concentrated on the daily management of pipeline, in order to edit water-supplying and water-draining graphic data and attribute data input, to query the attributes of pipeline and various attached devices and the analysis with certain traits and quantity, to realize the display and output of various images(including vertical and horizontal transect image of pipeline) and statistics, to maintain pipeline topology structure and update dynamically, so that it can realize the informative management of water-supplying and water-draining pipeline. The integration of pipeline graphic warehouse, attribute database and external database formed by GIS data integration facilitates dynamic update, so that the efficiency and quality of pipeline management could get improved greatly.

3.2 Model analysis of water-supplying and water-draining pipeline

The application of GIS in water-supplying and water-draining industry can provide decision support for pipeline planning and design, construction, as well as refinement schedule. It mainly includes pipeline model, GIS refinement design and graphics management.

Pipeline executes water force analysis for pipeline system, its basic data if from the analysis GIS rebuilding(expanding) model laid on pipeline; refinement design executes planning and analysis on the basis of status analysis and GIS system; graphics management is used to add, delete GIS graphics, manage project construction graphics, pipeline model graphics and status analysis graphics; refinement schedule executes schedule in all areas by utilizing basic data, current status data, GIS graphics data, SCADA real-measured data, water magnitude data, and delivers schedule orders.

3.3 Water-supplying and water-draining pipeline emergences pre-warning analysis

The features of city's water-supplying and water-draining pipeline would be demolished by various faults, to ensure the reliability of water-supplying (water-draining), the key point is how to find out the position of accidents fast and exactly, adjust valve in time, make the repairing time the shortest, waterless region the smallest and closed valves the least.

The traditional processing scheme is described on the paper by technicians, but this usually can't provide topology structure information of pipeline punctually and exactly, it doesn't just only waste a lot of man power, but also may inflict decisions fault. If explosion happens at some place in the pipeline, users should call the tap water company or some related department first, then the system operator will locate relevant graphic files in pipeline maps and attributes database according to the block name users informed, and determine pipe number, pipe diameter, pipe material, buried time and exact geography position, then run the valve schedule program, which results in the scheme of valve refinement schedule, after all, print these data out and hand them over to executers, so that you can prepare materials and get into scene and do the repair.

3.4 Water-supplying and water-draining pipeline patrol-along-line maintaining management

Using GPS technology to execute patrol-along-line is a trend of current water-supplying and water-draining industry. With the GIS platform's support, GIS patrol-along-line feature can use Unicom CDMA1X network and manage water-draining pipeline by GPS one position service and network technology. So as to realize communication between patroller and other units, and the results can be transfer to related department swiftly, so that each department can react to the feedback quickly. At the same time, can provide verbose, complete data for decisions, in order to manage those conditions more scientifically.

On on-hand terminals, you can connect graphic information with device information through GIS technology to accomplish "picture-number", "number-picture" query, and integrate graphic management with credit management together. Execute editing actions on graphics and database; swiftly locate relevant position on maps by getting position information through network; play back the patrolling trace of patroller according time periods; summarize according the feedback returned by field crews, generate statistics. Do regular patrol, special patrol and check and accept patrol. Meanwhile, connecting on-hand terminal with computers through USB interface enables you to synchronize the information of on-hand terminals and GIS system, by the upload and download features of GIS system. So that it realize the communication between patroller and individual unites in real-time, and the patrolling information will be transferred to all department quickly, in order to allow each department to react to the feedback quickly. And provide complete and verbose data to leader's decisions, in order to realize scientific management for patrolling.

3.5 Water-supplying and water-draining industry information incorporating management

As the science and city's development, the scale of water-draining pipeline is also expanding, rationally using high-tech to guide and help producing, realize information formation is the first priority for water-draining industry construction. By using GIS technology and combining with the water-supplying and water-draining industry's producing business process, you can manage water-producing techniques, water-supplying schedule, pipeline GIS, business charge, costumer service and engineering devices in a incorporative scheme, and can build enterprise synthesis informative platform, realize data integration management, sharing and digging, and finally realize society-oriented, leader-oriented, decision-oriented synthesis service system. The build-up of enterprise information platform enables individual business system to break the old running scheme of "do for himself", to realize complete resource sharing and application integration from data to functions. At the same time, provide work automation function, such as, network meeting, resource pre-order, file transferring, mutual data transferring and calendar warning. Decrease redundant effort and lower running cost. Prepare the backup unitively and improve efficiency.

In 2002, as the example project of national "ten-five" important project subject "city planning public business management system", Shaoxing tap water company cooperated with Wuhan zhongdi and started to build "Shaoxing tap water company water-supplying synthesis information system", which forms water-supplying synthesis information platform. This system use the integrate information of water-supplying business as a start point to realize complete integration and sharing for information by using new-completed water-supplying schedule, business management and pipeline geographic information as the basic structure, which enables the managing work of Shaoxing tap water company to jump to a brand new level, and fully improved its automatic level of water-supplying schedule, business management and pipeline management.

4 Case study, MAPGIS water-supplying and water-draining pipeline geographic information system

For the water-supplying and water-draining pipeline geographic information system built up based on domestic platform MAPGIS, it uses some newest technologies, such as GIS technology, communication technology, database technology and network technology, to realize water-supplying and water-draining information unitive management and data sharing, and on the basis of combining the business process of water-supplying and water-draining management, to realize the automation, science and standardization of information management.

(1)The spatial database of GIS manages the spatial deployment of terrain graphics, tour graphics, administration graphics, water-supplying pipeline, rain pipeline, sewage pipeline and various facilities, so that you can look up and output all information of water-supplying and water-draining pipeline at any time. This can fully reflect a whole appearance of water-supplying and water-draining pipeline, and also can look into every detail, which results in realizing the most complete and downright management for water-supplying and water-draining pipeline.

(2)Realize the dynamic management of urban water-supplying and water-draining pipeline basic data. The system can not only manage pipeline data been transferred, but also can add the pipeline data of new-built and rebuilt project to pipeline database, in order to realize the synchronous update of pipeline data, and to ensure the topology integrity of pipeline spatial data and veracity of attribute data.

(3)Can exactly retrieve lookup and output the graphics, technical data and forms that the water-supplying and water-draining projects need, and provide basic data support for the planning, design and construction of water-supplying and water-draining pipeline, improve projects' design and construction efficiency and reduce investment.

(4) Through water-supplying and water-draining expertise background and the space and network analyzing ability of GIS, it can execute professional analysis (such as pipe explosion analysis, water-draining area analysis, recurring period analysis, water-overflowing range analysis, pump station influence analysis), so that it provides a solid foundation for the application of water-supplying and water-draining pipeline on higher level.

(5) Helps you take care of accidents quickly. When there is an accident with water-draining pipeline, such as draining exceeded standards, water soiled, pipe is broken, system can find out accident spot, make repairing scheme and inform related units by draining user management, exceeding user tracking, draining area analysis, recurring period analysis, accident processing models.

(6) Evaluate the pipeline status, and calculate workload of daily patrolling, maintaining for pipeline, constitute plans, execute verification, in order to improve working efficiency and managing quality.

(7) Water force calculation. System can abstract existing data automatic all to calculate the water force of water-supplying and water-draining pipeline.

(8) Realize the publication of pipeline information on internet and LAN.

(9) Establish interfaces with other systems, such as schedule (SCADA) system, charging system, and realize system integration and data sharing.

5 Ending words

The prospect of the application of GIS in water-supplying and water-draining industries is very vast, it's crucial for most of the users to choose a mature, excellent GIS platform according the development of GIS technology. Nowadays, a lot of excellent GIS software was published internationally, but it's obvious that using domestic GIS software has so many advantages, domestic GIS software has good performance but sells in lower price, and it also has Chinese character feature and it's easier to be maintained and updated. In recent domestic GIS software, the MAPGIS, developed by China geography university (Wuhan) is famous, it's either good at 3S technologies(GIS , RS, GPS) incorporation or graphics processing, space analysis and interface, and it's picked by many users to be basic platform on which the second development is executed, and becoming more and more perfect in the process of users' utilizing and developers' research.

As the computer technologies and GIS technologies are developing and application request is getting deeper, the following aspects are obtaining more concerns.

5.1 Combined with internet

Internet is becoming the most important part of people's regular life and work. The pipeline information system adopting internet technologies can expand. The distribution area of workstations effectively, this facilitates citizens' query.

5.2 Combined with daily work more tightly

Pipeline information system needs to integrate various data which is under working environment, introduce the conception of workflow into pipeline management; this allows all working components associated with pipeline to become a whole entity.

5.3 Develop to synthesis integration

Pipeline information system will vastly integrate GIS technologies, GPS, multi-media technologies,

CAD technologies and AI technologies to develop the collection, management, synthesis analysis and process of urban water-supplying and water-draining pipeline to a more powerful spatial decision support system or a professional system.

Reference

[1] Chen Shupeng, Lu Xuejun, Zhou Chenghu. Geographic information system guide, published by science publishing company(Beijing),1999, 2-8.

[2] Zeng Wen, zhang Dejing. The design of MAPGIS pipeline development platform [J]. earth science—news paper of China geography university , 2002, 27(3): 250-253.

[3] Xu Shirong, Qiu Zhenghua. The calculating theory and electronic application of water-supplying pipeline. hunan university publishing company(Changsha), 1999, 6.

[4] Zeng Wen, Qiao wei. Using MAPGIS to develop urban water-supplying and water-draining pipeline information system, paper collection of Chinese GIS association meeting, 1999.

[5] Zhong Di. The water-supplying pipeline information system development report for Shaoxing, 2000, 10.

[6] Zhong Di. The water-supplying pipeline information system development report for Shuzhou, 2000, 12.

[7] Qu Fubang. Urban underground pipeline investigation research and application, southeast university publishing company, 1998, 12.

岩溶地下水污染控制技术

马振民 于玮玮

(济南大学城市发展学院，山东，济南 250002)

1 概述

凡是在人类活动影响下，地下水水质朝着恶化方向发展的现象统称为"地下水污染"。

人类活动导致进入地下水并使水质恶化的溶解物或悬浮物，无论其浓度是否达到使水质明显恶化的程度，均称为地下水污染物。它分为化学污染物、放射性污染物和生物污染物三类；来源按成因分为人为污染源、天然污染源；按分布形式分为点污染源、面污染源、线污染源；地下水污染途径有间歇入渗型、连续入渗型、越流型及径流型。

地下水系统按照系统结构、含水介质特征、水动力特征及边界性质等分为孔隙水系统、裂隙水系统及岩溶水系统。中国北方岩溶水系统地下水已受到不同程度的污染，局部地区岩溶水已失去其使用价值。

2 地下水系统污染源控制

在进行污染地下水治理之前，必须控制污染源。如果污染源得不到控制，污染物仍源源不断地进入包气带及其下的地下水，污染控制与治理就不可能取得成功。

2.1 不可清除的污染源的控制

城市垃圾、工业垃圾及放射性废物等固体废物，在目前的科学技术水平下，这些废物还不能完全消除。为了使进入包气带和地下水的污染物减少到最低限度，必须采取控制措施。

2.2 可清除的污染源的控制

抛撒在地面的贮存罐、贮存箱、废油桶等废设备，还有污水渗坑、排污渠道以及破损渗漏的设备等都属于可清除的污染源。其控制措施一般是停止使用，或迁移到安全的地点。

3 岩溶水系统地下水污染控制及治理技术

3.1 工程生物恢复技术

3.1.1 基本概念

工程生物恢复技术是指利用微生物，将地下水中的有毒有害的有机污染物转化为无害物质，或完全降解为 CO_2 和 H_2O 的方法，称为生物恢复。

生物恢复技术的优点是费用低，环境影响小，因为它的最终产物是 CO_2、H_2O 和脂肪酸；处理水平较高；可用于技术上难以应用的场地。缺点是不能降解所有的有机污染物；岩溶介质渗透系数太低时微生物生长可能产生堵塞等。

3.1.2 用于生物恢复的微生物

用于生物恢复的微生物有土著微生物、外来微生物及基因工程菌。大部分是应用土著微生物，外来微生物及基因工程菌适于降解专一的有机污染物。

3.1.3 生物恢复技术

根据所应用微生物种类的不同，生物恢复技术可分为两类：

(1)强化生物降解或生物刺激技术。通过引入无机营养物(主要是 N 和 P)、电子受体或有机物到地下水环境里,改变场地的物理、化学条件,增加土著微生物的活性和降解能力,加强其对污染物的生物降解。

(2)生物强化技术。把经过筛选和培养的微生物引入地下环境,以增强对特定有机污染物的降解。

3.2 水力控制技术

当地下水污染状况查明后,为了使污染的地下水不扩散到未污染区,最简便的处理技术就是水力控制技术。可分为三种系统:抽出-处理系统、水力隔离系统和非水溶性液体回收系统。

3.2.1 抽出-处理系统

抽出-处理系统的基本运转程序是通过置于污染羽状体下游的抽水井,抽出已污染的地下水,再通过地上的处理设施,将溶解于水中的污染物去除,最终,把净化水排入地表水体、回用或回注地下补给地下水。

此系统的优点:简便有效,对于泄漏发生初期污染程度高时效率高、应急。缺点:残留在多孔介质中的 DNAPLS 难以被抽出;污染物浓度较低时,耗时长、效率低、成本高,难以达到处理目标,对 DNAPLS 无效。

3.2.2 水力隔离系统

水力隔离系统也称为污染羽状持久稳定系统。其类似于抽出-处理系统,但它只起到稳定地下水污染羽状体,使它不扩延至未污水体的作用,并不能使地下水恢复其原来的环境功能。方法是,在污染羽状体前缘设置一个或多个抽水井,一排或多排抽水井,以形成污染地下水与未污染地下水间的一个水力隔离带,目的是保护未污染的地下水。

该系统使用条件:污染源位置无法确定时;由于人力或财力不足而暂时无法清理污染源时;有沉到含水层底部或深处的 DNAPLS 类污染物存在,抽出-处理系统处理无效时。

3.2.3 非水溶性液体回收系统

非水溶性液体回收系统是指比水轻的非水溶性液体,由于贮存罐的渗漏或偶然事故而渗入地下,在潜水面上形成一层漂浮的非水溶性液体。回收后再处理水中的溶解部分。回收的方法有井回收和截流沟回收。

3.3 治理地下水有机污染的气提技术和曝气技术

溢漏到含水层中的石油烃多以溶解态、吸附态和残余液态等三种基本形态存在,溶解态的石油烃进入地下水中,吸附态的石油烃吸附在含水层介质颗粒表面,残余液态的石油烃被截流在部分空隙里。转入溶解态的小于 5%,绝大部分是吸附态的和残余液态的。

针对地下水系统挥发性有机污染物,提出了气提技术及曝气技术(两者的区别在于挥发作用还是生物降解作用为主)。

3.3.1 基本原理

(1)污染物挥发。一是含水层介质中吸附态和残余液态有机污染物的挥发;二是地下水中溶解的有机污染物产生气提。当气流驱替地下水后,造成原来饱水的含水层暂时充满了空气,使吸附态和残余液态的有机污染物暴露在曝气环境里。此时,其蒸气压大于 133 Pa 的挥发性有机污染物都可能挥发成气态,随气流进入包气带,再通过气体抽吸井截获。此外,压力气流进入地下水可使溶解于地下水的有机污染物通过气提而挥发。

(2)生物降解。天然条件下,空气与水面接触,空气中的氧扩散进入水中,由于过程缓慢很难促进好氧生物降解。采用地下水曝气技术,井中和含水层介质中的水完全被驱替,空气暂时充满含水层介质的空隙,并溶于薄膜水中并使其充氧,缩短了空气中氧的扩散途径,因此有机物的好氧降解明显增加。

(3)有机物溶解增加。空气通过含水层介质引起介质空隙的扰动,使得部分被吸附的有机污染物和残余的有机液体与水混合,使分配到水中的有机污染物增加,溶解态的有机污染物更易于生物降解。

3.3.2 应用地下曝气技术的条件

(1)污染物类型。污染物是较易挥发($KH>10^{-5}$ atm·m^3/mol)且能被好氧生物降解的;挥发性差但可被好氧生物降解的,或者难生物降解但挥发性较高的。

(2)水文地质特征。处理场地的水文地质特征直接影响曝气技术应用的有效性。曝气技术是以空气注入饱水介质为基础的,注入的空气在水平和垂直方向上和通过饱水介质和非饱水介质流动,任何阻碍气流的因素都将影响曝气技术的应用,最理想的应用条件是比较均质的地层结构;适于曝气技术的地层应有较高透气性($K>10^{-7}$ m/s),地层的透气性应使空气在垂直和水平方向上通过包气带和含水层。

3.4 地下水污染的天然衰减恢复技术(MNA)——天然生物恢复技术

当含水层具有一定的天然容量,可以使地下水中的污染物发生生物降解、吸附、挥发、扩散等作用时,可采用地下水污染的天然衰减恢复技术,该技术是指没有人为干预,地下水中的污染物的总量、毒性、迁移性、体积和浓度在污染场地天然条件下各种物理、化学、生物作用过程中得以减少,使污染场地的地下水环境功能得以恢复。此技术主要工作在于监测、模拟、评价,证实污染物生物降解、稀释等衰减作用的存在,研究污染物的衰减速率和迁移速度。

3.5 地下水石油烃污染的天然生物恢复技术

溶解的石油烃是地下水最普遍的有机污染物。它们来自燃料、溶剂和木材防护剂的渗漏,以及含石油烃的各种废物的排放。

3.5.1 地下水系统中石油烃类的生物降解反应

地下水环境中石油烃的生物降解是在微生物参与下的氧化-还原反应。石油烃被氧化,给出电子(电子供体);电子受体(如 O_2 和 NO_3^- 等)被还原,接受电子。在地下水系统中,能作为电子受体的有溶解氧(O_2)、硝酸盐(NO_3^-)、Fe(Ⅲ)(如 Fe^{3+}、$Fe(OH)_3$ 等)、硫酸盐(SO_4^{2-})和二氧化碳(CO_2)。地下水中烃类的生物降解速率主要取决于三个因素:营养及电子受体的种类和数量;微生物的种类、数量及其代谢能力;烃的成分和数量。

3.5.2 地下水系统中降解石油烃的土著微生物

在很深的地下水环境中也存在着各种微生物群落,它们以细菌为主。天然的地下水环境往往是低营养环境,土著微生物对低营养环境有很强的适应性。许多微生物在高营养环境条件下繁殖很差或根本不繁殖,但在低有机碳环境下繁殖旺盛。

在天然的地下水环境中已发现 100 属 200 多种降解石油烃的微生物。微生物通过自然突变形成新的突变种,也可以通过基因调控产生诱导酶而适应新的环境。产生了新酶体系的微生物,就具备了新的代谢功能,从而降解或转化那些不能降解和不能转化的有机污染物。因此,本来对微生物有毒或抑制其生长的有机污染物,就变成了微生物生长不可缺少的营养物。

研究证明,受石油烃污染的地下水中有大量的降解石油烃的细菌存在,是地下水石油烃污染天然生物恢复技术应用的先决条件。

3.5.3 地下水石油烃污染的天然生物恢复技术的实施

天然生物恢复技术往往与抽出-处理技术联合使用。当发现地下水受石油烃污染时,首先采用抽出-处理技术,降低污染物的浓度,但要达到理想的处理目标往往要运转几年甚至几十年。在此情况下,应实施天然生物恢复技术。

石油烃污染的地下水是否可通过天然生物恢复而使污染物得到完全(或大部分)去除需要做许多工作,包括场地环境条件调研、系统监测和预测。

场地环境条件调研包括污染物是否可生物降解、污染物在含水层中是否生物降解及环境条件对生物降解是否合适;系统监测包括内层监测井和外层监测井烃类污染物和生物降解的指示参数

(如 DO、NO_3^-、SO_4^{2-}、HCO_3^-、Fe、Eh、CO_2、CH_4 等)监测，目的是判断污染是否向外扩散及烃类浓度的变化；预测必须取得污染羽状体的迁移速度和天然生物降解速率。迁移速度通过水质模型求得。天然生物降解速率取得有三种方法，即根据污染源下游不同距离监测井污染物浓度变化估算、室内微环境试验求得以及现场试验求得。

3.6 强化生物降解技术(生物刺激技术)

此技术采用循环井及格栅技术实现。针对地下水中土著微生物数量少和活性不高的特点，通过人为加入某些代谢基质(第一基质、电子供体)，刺激微生物的活性增强和数量增加，提高有机污染物的共代谢降解效率，从而达到对污染物的有效去除和地下水环境的恢复。

3.7 渗透反应格栅(PRB)技术

3.7.1 基本原理

渗透反应格栅(简称 PRB)技术是一种地下水污染的就地恢复技术，也可作为污染地下水的地面处理设施。PRB 由透水的反应介质组成，置于污染羽状体的下游，通常与地下水流相垂直。污染地下水通过 PRB 时，产生沉淀、吸附、氧化-还原和生物降解等反应，使水中污染物得以去除，在 PRB 下游流出达到处理标准的净化水。目前主要有以金属为反应介质的氧化-还原反应格栅及以释氧化合物为介质的生物降解反应格栅。

3.7.2 渗透反应格栅的种类

(1)按反应性质分，可分为化学沉淀反应格栅、氧化-还原反应格栅、吸附反应格栅及生物降解反应格栅。

(2)按结构形式分，可分为隔水漏斗-导水门式格栅、连续墙式格栅及灌注处理带式格栅。

3.7.3 渗透反应格栅设计中几个重要问题

(1)处理场地特征的调查。包括地质结构、含水层类型、水化学参数、污染物的种类和浓度、污染羽状体的范围及形状等的调查。

(2)反应系统处理能力的研究。在目前 PRB 发展阶段，不能提出标准的处理能力的设计方案。要确定反应系统处理能力，必须进行批试验、实验室或现场的渗流柱(土柱)试验。其目的是了解反应过程产物、污染物的半衰期和反应速率、反应动力学方程、等温吸附方程及分配系数、影响反应的某些地球化学因素等。

(3)反应系统停留时间的确定。反应系统的停留时间 t 设计是否合理既涉及到系统出水是否达到处理目标，也涉及到系统的经济合理性。如 t 太短，则出水可能达不到预期目标；如 t 过长，则会使格栅厚度过大，造成经济上浪费。

城市雨洪水利用与回补岩溶地下水

王维平　曲士松　邢立亭　孙小滨

(济南大学城市发展学院，山东，济南　250002)

降雨是自然界水循环系统中的重要环节，雨水对调节、补充地区水资源和生态环境起着极为关键的作用。一方面许多城市面临缺水、洪涝灾害、水污染及生态环境恶化；另一方面城市不透水面积的增大，改变了天然水文循环过程，城市为了防洪不得不通过建设庞大的排水系统将雨水径流尽快排出，造成雨水资源的浪费。中国北方地区城市水资源的短缺日趋严重，人类活动对地表径流和地下径流的影响加剧，进一步减少了可利用水资源量。雨水直接利用和利用雨水补给地下水已成为一种城市有效的淡水资源利用措施。雨水属轻度污染水，有机污染物含量低，溶解氧接近饱和，钙质含量低，总硬度小，易于处理，处理后可用做饮用水、生活杂用水、工业用水等。雨水利用和回灌地下水比处理后的生活污水回用更便宜，且工艺流程简单，水质更可靠，细菌和病毒的污染率低，经处理或净化的雨水其公众可接受性强。因此，尽管大气污染、下垫面材料影响雨水的水质，但只要处理好初期雨水，雨水水质还是较稳定的，是地下水回灌的可靠水源。因此，开展城市雨水利用和在城市开发建设区利用雨水回灌地下水对城市水资源可持续利用和水生态修复具有重要的作用和巨大的经济、社会、生态效益。

1　城市雨水利用与人工地下水回灌研究动态

城市雨水的利用是从 20 世纪 80~90 年代发展起来的。它主要是随着城市化带来的水资源短缺、环境和生态问题而引起人们的重视。许多国家开展了相关的研究并建立了一批不同规模的示范工程。在此基础上，城市雨水利用首先在发达国家逐步进入到标准化和产业化的阶段。在技术上领先的国家已进入标准化和产业化的阶段。例如德国在 1989 年就出台了雨水利用设施标准(DIN1989)，对住宅、商业区与工业区雨水利用设施的设计、施工和运行管理作出了详细的要求，并在过滤、储存、控制与监测四个方面制定了标准。到 1992 年已出现"第二代"雨水利用技术。又经过 10 年的应用与完善，发展到今天的"第三代"雨水利用技术。德国在雨水综合利用方面的研究始终位于世界科技的前沿，在 1990 年就已经发布了"对未受污染雨水的分散回灌系统的建设和测量"，1999 年又发布了"雨水回灌系统的设计、施工、运行"规范，将雨水回灌加以法律保护。目前德国在新建小区(无论是工业、商业、居民区)均要设计雨洪利用项目,否则政府将征收雨洪排放设施费和雨洪排放费。日本利用雨水作为生活杂用水的技术已经比较成熟，正大力实施雨水回灌以补充地下水。澳大利亚、德国、印度、法国和日本等国都在实施地下水人工补给，以解决国内水资源短缺问题。瑞典、荷兰和德国的人工补给含水层工程，在总供水中所占的份额分别达到 20%、15%和10%。美国正在实施"含水层储存回采 ASR 工程计划"，到 2002 年 7 月，正在运行的 ASR 系统共有 56 个，而建成的系统则有 100 个以上。

中国城市雨水利用起步较晚，目前主要在缺水地区有一些小型、局部的非标准型应用。比较典型的有山东的长岛县、大连的獐子岛和浙江省舟山市葫芦岛等雨水集流利用工程。大中城市的雨水利用基本处于探索与研究阶段，但已显示出良好的发展势头。北京、上海、南京、大连、哈尔滨、西安等许多城市相继开展了雨水收集利用项目研究。由于缺水形势严峻，北京在雨水收集利用技术方面走在了全国的前列。北京市政府 2000 年 66 号令"城镇地区机关、企事业单位院内应当建设雨水收集利用的设施，鼓励单位和居民庭院建设雨水利用设施和渗水井"，北京市雨水利用已进入示范与实践阶段。20世纪60~70年代，人工回灌曾在我国风行一时，主要是为了补给地下水，缓解供水紧张，同时也是东南沿海城市防止地面沉降和海水入侵的主要措施。没有得到广泛推广的主要原因是注水井常常发生淤填。进入90年代，中德合作项目在北京市采用土壤含水层处理系统(Soil Aquifer Treatment)，对经过深度处理后的污水进行人工地下水回灌示范研究。2001年9月中德科技合作项目在北京开展了"城区水资源可持续利用——雨洪控制和地下水回灌"，中德合作项目在北京市采用渗井，开展了在现存和新开发城市区域通过搜集、存储和处理雨洪水并回灌浅层地下水示范工程。上海市近二三十年来一直采取压缩开采量、增加回灌量的方法来控制地面沉降。沈阳市1987~1991年开展了地下水人工回灌试验及其预测的研究。"七五"期间，中国水利水电科学研究院在北京大兴县开展了注水井回灌农业灌溉用地下水的研究，有一定效果。2002年山东省鲁北地质勘察院进行了德城区深层地下水人工回灌试验调查项目。

到目前为止，在半湿润半干旱地区的城市开发建设项目区，如何利用雨水人工回灌岩溶地下水这方面的研究，国内外很少报道。由于涉及到雨水的水质和水量、建设开发项目下垫面对雨水水质水量的影响、岩溶地区水文地质条件、雨水储蓄设施、回灌井等多方面内容，利用雨水回灌岩溶地下水有许多问题需要进行深入研究，以便为制定相应的技术指南、标准和规范提供依据。本文以济南市为例探讨雨水利用及回灌岩溶地下水的途径。

人工回灌在最近二十几年里才取得了长足的发展，在提供饮用水和进行高标准废水处理方面积累了很多经验。一般而言，人工回灌地下水主要有两种方式：一种是在透水性较好的土层上修建沟、渠、塘等蓄水建筑物，利用水的自重进行回灌；另一种方式就是井灌。前者是人工回灌的最简单形式，后者适合于地表土层透水性较差，或地价昂贵，没有大片的土地用于蓄水，或要回灌承压含水层的情况。现在对地下含水层的人工回灌和再利用，国际上通称为ASR技术或ASTR技术，ASTR技术与ASR技术的区别在于前者的注水井与抽水井并不是同一口井，使得注入的水可以在含水层中运移过程中被进一步净化，这样抽出来的水就可达到饮用水的要求。其基本过程如下：收集(雨水、废水)→预处理→回灌→抽取→再利用(饮用、灌溉等)。通过透水性良好的土层人工回灌潜水含水层已有了很长的历史。Dillon等统计出当前10个国家中70多个人工回灌地下水的事例，回灌能力从每年几千立方米到几百万立方米不等，在有资料可查的40例中的38例在技术上是可行的。ASR技术中，最核心的问题是注水井的堵塞和保证地下水不被污染。

2 济南市城区雨水利用和回补地下水的必要性

济南市素以泉水著称，市区四大泉群通过护城河与大明湖形成了济南市美丽的水景，既是自然景观也是文化遗产。同时良好的地下含水层储存构造，为城市居民生活和工业提供了优质的岩溶地下水。然而，由于岩溶地下水超采，泉水在20世纪80~90年代经常停喷。90年代以来加上城区面积迅速增加，特别是城市向南部地下水直接补给区扩展，大量硬化面积使得地下水入渗补给减少，地表径流增加，下游河道防洪压力愈来愈大。因此，如何利用雨水回灌岩溶地下水，是济南市供水保泉的关键措施之一。

2.1 雨水直接利用和入渗对济南市区防洪具有重要作用

随着济南市城市化的进程，不透水面积也相应增加。根据统计，目前济南市城区面积已由解放初期的26 km² 增加到2004年的278 km²。如果按不透水面积占新城区面积的70%计算，新增1 km² 城区面积，将会比自然条件新增地表径流量47万 m³。济南市区目前的排水系统主要利用天然的河道和冲沟，最终排入小清河。目前，城区河道防洪标准不足50年一遇，加上许多排水排洪河道被建筑物占用，垃圾堵塞，若对下游河道断面排水进行扩大，一是成本高，二是难度大，三是能力有限。因此，雨水利用和入渗是减轻济南市下游防洪压力的有效措施之一。

2.2 利用雨水补给岩溶地下水对保泉和供水具有重要作用

中国北方溶岩分布地区地下水多为裂隙岩溶水，裂隙岩溶水系统多以溶蚀裂隙为主组成独立的网络系统，赋存地下水层位属中奥陶统–寒武统，裂隙岩溶水已成为中国北方部分城市主要的供水水源，山东省城市供水的主要水源一半是地下水。全省石灰岩区面积 18 014 km^2，其中裸露区面积 15 059 km^2，涉及济南、淄博、泰安、枣庄、济宁、临沂六个地市，在这些地区岩溶地下水在城市供水中起着关键的作用。

济南市具有较好的岩溶水赋存条件和得天独厚的地下水补给条件，其中市区四大泉群的泉水喷涌量，主要来源于南部山区直接补给区内的降雨入渗补给。城区的发展，在直接补给区同样面积，建设后比建设前至少减少了 2/3 的降雨对地下水的入渗补给量，是影响泉水喷涌的最直接原因之一。过去城区盲目向南部山区发展，已建了大量的开发项目，目前尽管济南市城市发展总体的规划是南控，但不是禁止，在规划区的地下水直接补给区内仍然继续发展。因此，在直接补给区内对已建和规划项目采取工程措施，利用雨水补给地下水，充分利用该区域地下水补给的有利条件，是解决南控中城市发展与保泉的尖锐矛盾，达到生态与经济社会可持续发展最有效的措施之一。

其次，根据雨水水质取样分析，济南市雨水水质相对较好。工业区位于北部，南部为山区，是济南岩溶地下水的直接和间接补给区，夏秋季一般刮西南风，大气污染物较少，但是非降雨期屋面上积累的大气沉降物还是较严重的，只要解决汛前初期雨水，雨水水质可以得到较好的控制。

2.3 利用雨水具有巨大的经济效益和生态效益

利用雨水和进行地下水补给工程相对大型水源工程来讲投资规模小、投资分散、运行费低，生态效益更显著。北京市对城市雨水各方案设计的技术经济分析比较，认为雨水渗透方案综合效益最为明显。在条件成熟、资金充足时，雨水的中水利用可作为严重缺水地区解决水资源危机的途径。

3 城市雨水利用回灌岩溶地下水的规划目标和思路

济南市雨水利用回灌岩溶地下水的目标尽可能通过人工回灌工程将城区岩溶地下水直接补给区的降雨、入渗、径流和蒸发恢复到自然状态。不仅是将雨水作为一种可重复利用的水资源来满足生产和生活的需要，同时也要在人口密集、不透水面积日益增多的城市重新发挥雨水在天然水循环中所起的作用。

思路是雨水资源化，通过工程将雨水蓄积起来或回补地下水，不仅可以增加城市供水水源，同时还可减少城市径流量，减轻城市排洪设施压力，减少防洪投资和洪灾损失。雨水利用与回灌岩溶地下水对济南市可持续发展具有重要的作用。但是，由于雨水利用涉及城建、水利、地质等多部门，需要相应的政策和法规支持以及多专业跨学科的协作。2010 年前以试验研究和示范推广为主，在此基础上结合济南市水资源管理条例以及济南市总体规划中的南控方案，制定出相应强迫性的济南市雨水利用管理条例，对新建项目和已建项目，从雨水利用工程的投资、工程形式、工程建设、运行管理等进行规定。2020 年前，使直接补给区雨水回补岩溶地下水量达到岩溶水开采量的 5%。

4 工程措施、布局规划

雨水利用有多种方式，从济南市实际出发，在岩溶地下水直接补给区内重点实施雨水回补地下水、减少地表径流工程和雨水直接利用工程。

4.1 主要工程措施

4.1.1 雨水回灌岩溶地下水工程

结合不同开发建设项目区的水文地质结构，以回补岩溶地下水为目的，以建筑群和小区为对象，同规划设计的以及现有的社区雨水排水系统结合，并以深层渗透井建设为核心。将屋顶路面雨水，通过雨水预处理和调蓄设施，回灌注水井，补充岩溶水，超过渗透能力的雨水再进入市政雨水排水管网。

4.1.2 雨水就地入渗工程

利用人工渗透地面，以防洪减少地表径流为目的。例如多孔的嵌草砖、碎石地面、透水性混凝土等以及天然渗透地面，将雨水渗入地下。天然渗透地面绿地一般具有较好的渗水性，在充分利用绿地良好的自然渗透和保水性的同时，利用人工促渗设施，增大绿地的入渗水量。将直接补给区内目前凸式绿地改成低凹式绿地，特别是规划的或现有的公园、草坪、房前屋后等绿地，设计或改造成良好的渗水场，在低洼处建设铺有滤料的渗渠、渗沟。

4.1.3 屋面雨水集蓄利用系统

以雨水直接利用为目的，以单独建筑物或建筑群为对象，屋面雨水经雨水竖管进入初期弃流装置，经处理后的雨水通过储存池储存调蓄，再通过配水系统，用做各种生活杂用水，包括浇灌、冲厕、洗衣、冷却循环等中水系统。

4.2 雨水利用工程布局和实施步骤

雨水利用工程主要应用于南部已建生活小区、城市总体规划的南控的适宜开发生活小区、政务区。工程形式和布局主要根据当地第四系覆盖层的厚度以及下伏石灰岩的特性进行优化布局。采用分步实施的方法，第一步，先启动雨水直接利用工程和雨水就地渗漏工程；第二步，综合雨水补给岩溶地下水工程，由于涉及到水质控制与回补效果，可先作雨水回补岩溶水区域评价，在此基础上，进行不同水文地质条件深层渗井示范工程试验，制定相应的技术标准；第三步，推广实施，到2020年，建成深层渗井100眼，回补岩溶地下水量500万 m^3，屋面雨水积蓄利用系统20个，直接利用雨水10万 m^3。

4.3 工程投资估算

2010年前，研究示范工程需投资1 000万元，根据目前北京市等地区已做的工程，雨水利用一般1 m^3 水投资84元，2010～2020年需投资840万元；综合雨水收集渗透排水工程，预计2010～2020年需投资42 800万元。合计为43 640万元。

5 制定雨水利用及回灌岩溶水的政策和法规

目前我国城市雨水利用刚刚起步。2003年我国只有北京市制定了有关雨水利用的政策《关于加强建设工程用地内雨水资源利用的暂行规定》。济南市《水土保持管理办法》第十四条规定"在地下水资源主要补给区及其保护范围内的南部山区，应当限制开发建设。经批准开发建设的，其用地中硬化面积不得超过总建设用地面积的百分之三十"，第十五条规定"开发建设项目应当尽量减少地表硬化，广场、露天停车场、庭院、人行道、隔离带等应采取有利于雨洪入渗的措施"。但是办法对采用何种措施、技术标准，资金筹措和建后管理都没有规定。因此，对济南市尤其是南部山区，最紧迫的是要制定出有关法规，使已建和拟建的开发建设项目，在回补岩溶地下水、减少地表径流、减小下游防洪压力和灾害方面，达到"以工程建设后不减少岩溶地下水补给量，不增加建设区域内雨水径流量和外排水总量为标准"。同时增加由于建设项目减少地下入渗补给量补偿费和防洪费20元/m^2，用做雨水利用工程建设和运行管理费。

6 雨水回灌岩溶地下水需解决的关键技术问题

在岩溶地下水直接补给区利用回灌井补源，由于涉及到开发建设项目及城市供水水源地的供水安全，非常敏感而且影响巨大，必须做好扎实的前期研究工作，解决以下技术上的关键问题，为制定济南市雨水利用及回灌岩溶地下水的技术规程提供依据。

(1)确定汛期济南市雨水水质状况及变化规律；

(2)初期雨水处理及雨水过滤处理系统；

(3)雨水储蓄工程；

(4)岩溶回灌井的选址及水文地质结构；

(5)岩溶回灌井的结构和形式及岩溶含水层的储存能力；

(6)回灌雨水对岩溶地下水水质的影响评价；

(7)雨水中的悬浮物、空气、生物膜等对回灌井及周围含水层的堵塞及雨水与岩溶岩产生化学反应和生物活动沉淀造成的回灌井和周围含水层阻塞；

(8)雨水回灌岩溶地下水监测；

(9)雨水回灌岩溶地下水的法律和法规；

(10)雨水回灌岩溶地下水的风险评价。

7 结语

在济南市岩溶地下水直接补给区的开发建设项目区内，利用屋顶雨水径流回灌岩溶地下水，既可以利用碳酸盐岩渗透系数大和补给区与泉群距离近以及屋面雨水水质较好的优点，使得雨水径流及时补给岩溶地下水，起到供水保泉的作用，也可以协调过去或未来城市向南部山区扩展和生态环境保护与恢复的矛盾，具有重大的经济、环境效益和社会效益。

由于利用雨水回灌岩溶地下水，不仅技术而且政策涉及多方面部门，除了需要做大量研究和示范工程，还需要各部门的协调。一旦该项技术成熟后，就可全面实施，它将会对济南市泉水的保护与恢复起到重要的作用，同时对山东省城市开发建设区岩溶地下水保护和利用具有显著的示范和推广意义。

参 考 文 献

[1] 武晓峰, 唐杰. 地下水人工回灌与再利用[J]. 工程勘察，1998(4): 37-42.

[2] 丁昆仑. 人工回灌地下水的有效途径和方法探讨[J]. 中国农村水利水电技术, 1996, 1(2): 14-17.

[3] 杨维, 陈曦, 李晨. 回灌条件下地下水质模拟与预测[J]. 工程勘察, 2002 (4): 23-25.

[4] 朱桂娥, 薛禹群, 李勤奋, 等. 回灌条件下地下水的动态特征——以上海市浦西地区第Ⅱ承压含水层为例[J].水文地质与工程地质, 2001 (4): 67-70.

[5] 济南大学, 济南市水利局. 济南市地下水行动计划[R]. 2006.

[6] W. F. Geiger, Li Zifu, Liu Jiangsong. New Technologies for Water Management in Residential and Commercial Area[J]. Construction Science & Technology, 2003 (11): 1-7.

[7] W. F. Geiger. Sustainable Management for Flood Control and Groundwater Recharge in Beijing[C]. XI. International Conference on Rainwater Catchments System, IRCSA. August 25-29, 2003, Mexico City.

[8] 车武, 李俊奇. 从第十届国际雨水利用大会看城市雨水利用的现状与趋势[J]. 给水排水, 2002, 28(3): 12-14.

[9] 王惠贞, 车武, 胡家骏. 浅谈城市雨水渗透[J]. 给水排水, 2001, 27(2): 4-7.

[10] 车武, 李俊奇, 等. 现代城市雨水利用技术, 节水新技术与示范工程实例[M]. 北京: 中国建筑工业出版社, 2004.

[11] 曹秀芹, 车武. 城市屋面雨水收集利用系统方案设计分析[J]. 给水排水, 2002, 28(1): 13-15.

济南保泉对策研究

邢立亭

(济南大学城市发展学院，山东，济南 250002)

1 水环境问题现状

1.1 泉水断流

1936 年济南趵突泉水厂日均开采量 1.28 万 m³/d，至 20 世纪 60 年代初，市区地下水开采量小于 10 万 m³/d，平均水位在 31.54～30.72 m，泉流量 35.52 万～33.58 万 m³/d，60 年代末至 70 年代中期，地下水开采量由 10 万 m³/d 增加到 27 万 m³/d，年平均泉流量减少到 15.22 万 m³/d 左右，70 年代末至 80 年代初，市区地下水开采量由 27 万 m³/d 增加到 30 万 m³/d，泉流量由 15.22 万 m³/d 减少到 10.48 万 m³/d 左右，90 年代至 21 世纪初，泉流量急剧衰减，年平均流量小于 5 万 m³/d。

1.2 水位下降

在 20 世纪五六十年代，济南市区年地下水平均水位高于 30 m，由于人为因素干扰，随后地下水位持续下降，至 1990 年枯水期泉水位降至历史最低水平 20.8 m，截止 20 世纪 80 年代中期初步形成市区、西郊和东郊开采漏斗区，地下水降落漏斗内水位低于趵突泉泉水出流标高 26.9 m，以地下水位标高 27 m 计算多年平均市区漏斗面积在 40 km² 左右。

1.3 地表水污染严重

济南水环境污染十分严重，根据监测，共检出 76 种有机污染物。据 2002 年 6 月分析资料，超标项目有酚类、氨氮、溶解性铁，其中酚超标地面水Ⅳ类标准，氨氮远远超过Ⅴ类标准，水体有机污染严重，已丧失其功能用途；2002 年监测大明湖水质总氮、总磷、化学需氧量、生化需氧量、高锰酸盐指数及大肠菌群数年均值超过《地表水环境质量标准》Ⅴ类标准，呈重度富营养化状态。

2003 年 9 月泉水喷涌补给大明湖后，大明湖水质逐步改善。由于历史遗留原因，大明湖南岸每天有 7 000 m³ 未经任何处理的高浓度生活污水排入湖内，造成大明湖水质恶化。

1.4 地下水受到污染的严重威胁

济南地区断裂构造发育，灰岩含水层富水性不均，单井出水量在 100~12 000 m³/d 之间。近 20 年来受人类活动影响地下水质量逐渐恶化，表现为地下水矿化度、硬度、NO^{2-}、SO_4^{2-} 含量升高或超标，酚、氰化物、油类、重金属离子检出或超标。如 2004 年监测东郊地区地下水的矿化度是 1958 年的 2.09 倍，总硬度是 1.46 倍，Cl^- 是 3.35 倍、SO_4^{2-} 是 7.75 倍，但氯离子、硫酸盐、矿化度、总硬度的增幅比市区、西郊大，说明东郊工业区三废排放对岩溶地下水影响大于市区和西郊。

1.5 河流干涸，沟谷淤积，地下水的补给量减少

济南泉域在南部山区面积近千平方公里，地层为寒武系中下统灰岩、碎屑岩和泰山群变质岩，入渗条件差，河流、沟谷纵横，地表径流发育，地下水就地补给，汇于沟谷短途运移、排泄，大部分河流发源于此。河流表流汇集到玉符河、北大沙河，向下游径流，河水进入灰岩段渗漏补给泉域地下水，如 1963 年卧虎山水库向玉符河放水 10 186.10 万 m³。卧虎山、锦绣川等水库修建拦截上游地表径流后，近三十多年来，河道基本长年干涸，地表水对地下水的补给量减少，特别是 80 年代后期，随着卧虎山、锦绣川水库规划为向市区供水以后，源自间接补给区的地表水补给逐

渐减少，近 20 年来，玉符河河道基本长年干涸，统计计算，与 80 年代以前相比，1999～2002 年源自间接补给区的补给量减少 18 万 m^3/d。

另外，南部灰岩地区沟谷是地下水重要补给源，目前沟谷内垃圾、渣土堆积，耕种农田，修建道路、住房，许多原本渗漏的河道、沟谷严重淤积，入渗能力降低。南部山区沟谷的生态环境恶化，植被稀少、顺坡耕作、山石开采、垃圾渣土堆积、沟谷淤积，涵养水源条件变差，入渗能力降低。

1.6 城市扩展，地下水入渗补给面积减少

济南市依山傍水，南为连绵的石灰岩山区，北临黄河。因地势南高北低，大气降水在南部山区渗漏，沿倾斜方向北流，南部山区是济南的"生命之源"。早在 2003 年济南市政府就确定了"东拓、西进、南控、北跨"的城市规划建设总体思路，但是，目前在地下水补给区仍然继续规划建设，硬化面积继续扩大，如长清大学园区等，更为严重的问题还出现在渗漏河道铺膜防渗，如 2006 年东风水库防渗工程铺底、河道中衬砌，使得该强渗漏区将变成非渗漏区。又如位于孟家附近的奥体政务中心，作为 2009 年全运会的场址，2006 年开工建设，本区石灰岩入渗条件好，第四纪厚度薄，生态环境极其脆弱，下一步应采取可行的保护措施，维护孟家水库及河道的渗漏功能。

城镇化建设对泉水影响表现为城市面积增加，根据多时相遥感解译，2001 年城区面积比 20 世纪 50 年代扩展 175.6 km^2。城区向灰岩地区扩展，直接补给区面积减少，地面固化，诸多地段成为永久性不渗漏区，雨水进入防洪沟，而防洪沟淤积严重，地形坡降大，不能形成有效入渗。在多年平均降水年份，济南城区向补给区南部、西南、东南地下水补给区扩展，减少地下水补给量达 1 400 万 t/a，这在中国北方岩溶地区规划建设中需要引起高度重视。

2 产生水环境问题的原因

2.1 科学研究投入不足，造成认识不统一，水源地布局不合理

水源地大多围绕泉群分布，如济南市区普利门水厂、解放桥水厂、百货大楼水厂、东郊白泉附近杨家屯水厂等，远离泉群的水源地未得到充分利用，如长孝水源地、黄土崖水源地。

济南市集中开采水源地距离泉群附近，同时城区向地下水补给区南部、西南、东南扩展，建立大量自备井，大量开采岩溶水，改变了地下水天然流场，袭夺泉流量(见表 1)。可见，大量开采地下水特别在市区直接抽取地下水是影响泉水出流的主要原因。

表 1 地下水开采量与泉水动态对比表

时间	平均开采量(万 m^3/d)	泉流量(万 m^3/d)	泉水动态
20 世纪 50 年代末	5.97	31.0	常年喷涌
20 世纪 60 年代	13.20	34.2	常年喷涌
20 世纪 70 年代	44.30	14.3	间歇喷涌
20 世纪 80 年代至 21 世纪初	50~55	5.0	断流多于出流时间

另一方面，由于各研究部门的技术资料不能共享，缺乏相互交流，对济南水文地质条件认识不同或部门利益驱使，开采东郊还是开采西郊多年来一直处于争论过程中，造成政府决策无所适从。

第三，由于供水管网不配套，泉群外围存在大量自备井，截获泉水的补给量；同时由于水资源属性不清，应该关闭的自备井没有关闭，部分被关闭的自备井又缺乏科学依据，采取一刀切形式。如市区北园路以北的孔隙水开采井。

2.2 引黄投入与水量产出不匹配

为恢复泉水减少了地下水开采量，1988 年兴建引黄保泉工程，投资 2 亿元的引黄供水一期工

程通水,设计供水 20 万 m³/d,实际平均供水不足 5 万 m³/d。目前,设计供水能力 80 万 m³/d、投资 22 亿元的鹊山和玉清湖两大水库建成蓄水投产后,由于众多原因,黄河水厂供水量不稳定,水质差,玉清湖渗漏量严重,最大渗漏量大于 20 万 m³/d,远远达不到设计供水能力,平均供水量仅 30 万 m³/d 左右,只能靠开采岩溶水来维持供水,多年平均开采量 60 万 m³/d 左右。

2.3 城市规划布局存在缺陷,新城建设将加剧水环境恶化

2003 年 6 月省委召开常委(扩大)会议,确认了"东拓、西进、南控、北跨、中疏"的城市发展战略。2005 年提交的济南市城市总体规划(2005~2020 年)第 36 条指出,落实城市发展"南控"方针,划定"城市建设南部控制线"为:城区东南部以旅游东路南侧规划确定的建设用地南缘为界;南部以兴隆山山脊线经双尖山、至大王寨山山脊线为界;西南部沿五峰山路高校科技园南部山体山脊线至济菏高速公路为界,"对于涉及地下水水源涵养保护区内的城市建设活动,应进行科学分析论证,予以重点控制"。

从水文地质角度,"南控"对于保泉具有重要意义。但是,总体规划划定"城市建设南部控制线"以旅游东路南侧的建设用地南缘—兴隆山—双尖山—大王寨—沿五峰山路高校科技园南部山体山脊线—济菏高速公路为界,根据济南地区水文地质条件分析,"城市建设南部控制线"以北至山前属于地下水直接补给区,因此南控红线必须重新界定并向北移。同时第 97 条指出:泉源的保护主要指泉水补给区、重点渗漏带的保护。由此可见,从水文地质条件分析,第 36 条与第 97 条相互矛盾。

济南市城市总体规划(2005~2020 年)中规划布局的"东拓"、"西进"区域依然存在缺陷。"东拓"、"西进"形成的西部城区和东部产业带将对水环境造成巨大危害,如城关片区、平安片区、崮山片区、雪山片区、围子山片区、莲花山片区、章锦片区、彩石片区、唐冶片区均位于岩溶地下水的直接补给区,部分地段基岩裸露,如汉峪片区基岩裸露面积占 33%,章锦片区占 20.4%。西部城区和东部产业带规划区建设后,地面硬化,必然造成地下水入渗补给减少;另一方面污水长期排放、管网渗漏,易造成地下水污染。

大学园区的西片区和南片区位于北大沙河西侧山前地带,规划生活区、产业园区和综合性医院等。北大沙河河床附近和平安店断裂两侧局部砂卵砾石层与下伏奥陶纪灰岩直接接触,无黏土隔水层,形成"灰岩"天窗,是河水、污水渗漏区段;本区是济西水源地的重要补给区,建设后地面硬化,不仅造成济西桥子李水源地、冷庄水源地、古城水源地补给量减少,而污水排放和污水管网渗漏造成水质恶化已不可避免。即使通过处理污水零排放,但入渗补给减少已经不可逆转。

2.4 水资源管理问题突出

首先,管理体制不合理使水资源问题日益突出。目前,水资源的开发、利用与保护涉及到水利、农业、自来水、节水办、城建、环保、林业等多个部门,多龙治水,"城乡分割"、用水计划审批部门与节水、规划部门分离,市区与郊区水资源管理部门分离,使水资源管理陷入被动局面。

其次,在地下水理论研究方面,地质勘察部门具有无可比拟的技术优势和资料优势,但脱离地下水资源开发与管理等重要环节。

第三,水环境保护意识淡薄,受地质条件约束,灰岩区土壤稀薄,生态环境相当脆弱,南部山区的盲目开发,水土流失严重,蓄水能力变差,降低灰岩地区沟谷的渗漏能力。石灰岩地区垃圾堆积、污水排放、建养殖场构成地下水潜在的污染源。

第四,水环境保护措施不当,目前南部山区保护示范区建设均位于岩溶水间接补给区,即寒武纪地层和泰山群地层分布区,而地下水的直接补给区缺少相应保护工程。

2.5 水资源优化配置的格局尚未形成

城市基础设施建设滞后,加剧了水资源的紧缺。首先,供水管网不完善,工业与生活同一管道供水,分质供水难度大,即使在东部产业带和西部城区也未进行分质供水管网配置。目前,东郊工业用水大多采用地下水,同时地表水水库未充分发挥作用,引黄的实际供水能力远达不到设计供水目标,地表水置换地下水格局未全面实施。

另一方面，污水利用率低下，济南市区有先进污水处理厂，污水处理规模较小，同时污水资源化工作进展缓慢，污水及中水回用量不足市区总用水量的 30%，如污水处理厂处理后中水，大部分排入小清河，尚未充分利用。

第三，农业灌溉一直抽取地下水，依然采用大水漫灌的方式，跑冒滴漏现象严重，灌溉定额高达 400 m³，工业万元产值耗水量大，供水管道漏失率高，水资源浪费现象严重。

2.6 沟谷淤积、滥采、滥挖造成入渗能力降低

南部灰岩地区沟谷是地下水重要补给源，城区向南部、西南、东南扩展，目前沟谷内垃圾、渣土堆积，耕种农田，修建道路、住房，许多原本渗漏的河道、沟谷消失，严重淤积。此外，南部山区属水土流失严重地区，采石造成地貌景观、植被破坏，拦蓄、涵养水源能力大大降低，每次降雨，都不同程度地造成城区洪水泛滥，桥涵积水，部分地段积水深度大于 1 m，交通中断。

3 水资源开发与环境保护协调发展的措施与建议

以人为本，坚持开发与保护并重，通过开源、节流、涵养水源、回灌补源、水资源优化配置等措施，保证供水及泉水常年喷涌，促进和保障济南市经济社会的可持续发展，是济南水资源开发利用的总体思路。

3.1 分质供水

济南地区主要含水层为寒武-奥陶纪裂隙岩溶含水层，裂隙岩溶水是济南市供水水源及泉水水源。大气降水是地下水主要补给来源，根据降水分析，济南地区多次出现"四枯一丰"的降水系列，采用美国地质调查局(USGS)开发的三维有限差分地下水流数值模拟的软件 MODFLOW 计算。预报时采用 1999～2003 年降水量相对较小的系列，初始水位为 2005 年 6 月 1 日，目标时间确定为 2010 年 5 月 31 日。其中，预报期间内第 5 水文年降水量最大，为 985.6 mm，第 4 水文年降水量最小，为 370 mm。由于西郊与市区存在水力联系，考虑保护泉水，未来城市供水布局必须调整，需要减小西郊水源地的开采规模，据模拟优化计算，正常年份控制泉水位标高在 28.5 m 以上，不包括农业开采，泉域岩溶水允许开采量 18.8 万 m³/d，其中，济西水源地开采 10 万 m³/d，西郊水厂开采 5.8 万 m³/d，东郊工业自备井控制在 3.0 万 m³/d；泉水先观后用 5.0 万 m³/d；虽然西郊有腊山、峨眉山、大杨庄、古城、桥子李和冷庄六个水厂，供水能力大于 46 万 m³/d，只能作为必要时应急供水。

由于保泉约束条件下优质地下水资源有限，建议分质供水原则：工业用水采用黄河地表水，农业在节水前提下限量开采地下水，中水用于生态用水，优质地下水作为生活用水。根据计算，白泉水源地可供水量 28 万 m³/d，长孝水源地建议开采 8 万 m³/d，那么济南泉域、白泉和长孝水源地合计优质地下水资源可开采量 55 万 m³/d，可用于生活和高精尖工业用水，优质岩溶地下水资源可用于 450 万人生活用水(规划 2020 年中心市区人口 430 万人)。

3.2 加强科学研究，统一认识

20 多年来，专家们一直在泉域边界这个关键问题上争论不休。水文地质条件固有其复杂性，但是新技术方法、勘查手段的缺乏造成对地下水补给、径流、排泄条件认识不一致，应加大经费投入，重点查明存在争议地段的水动力场，只有查明地下水补给来源，保泉才能有的放矢，同时为城市规划布局提供依据。

3.3 加大执法力度，关停自备井，健全立法工作，保护山区生态

在完善供水管网前提下，加大执法力度，彻底关停岩溶水自备井，鼓励开采沿黄浅层地下水；从水文地质角度划分地下水易污防护区，保护南部石灰岩山区生态环境，完善法律法规，禁止滥采滥挖，开山采石，破坏植被，逐步改变南部山区产业结构，涵养水源。

南部山区作为水源涵养生态功能区，生态环境保护与建设重点是通过退耕还林、荒山绿化、小流域治理、矿山开采区恢复治理、水资源调控、治理水土流失，以提高水资源涵养能力，加强

生物多样性保护，防治废水和固体废物污染以及农业面源污染，建成为"山、水、绿、人"和谐统一的可持续发展区域。

3.4 回灌补源，进行地下水系统功能修复

泉水断流、水质恶化是人类活动影响的地下水系统功能的退化，必须进行地下水系统功能修复。南部山区石灰岩岩溶发育，具有极强的渗透能力。利用河流、沟谷进行人工补源，可以使山前地带及市区地带地下水位抬高，2003年玉符河回灌试验表明，玉符河渗漏补充地下水，大幅抬升西郊地下水位。实践证实，人工补源可以使地下水系统增源增采，因此在泉水补给区的渗漏地段，建设拦洪蓄水工程，拦蓄地表水或引水渗漏补源，可提高市区地下水位，这是未来保泉的根本措施。

根据调研，目前直接补给区缺少拦蓄工程，而在间接补给区却修建拦蓄工程，不仅达不到抬升市区水位的作用，相反增加蒸发，浪费了资金。

3.5 加强部门协作，科学保泉

保护与恢复泉水涉及多学科，需要多部门联合协作，从地下水补给来源和排泄去路两方面做工作，如在地矿部门查明地质条件，查明地下水运移通道和渗漏段基础上，水利部门实施水利工程调水补源，林业部门在直接补给区内增加植被覆盖率，环保部门治污重点区应该在地下水直接补给区，而不是在火成岩地区。多部门协作科学保泉，可大大减少工作的盲目性、片面性，同时也可节约资金的投入。

3.6 控制城区向直接补给区内扩展，以免使泉水补给量减少和造成水质污染

通过2004年10月31日至2005年9月30日一个水文年的资料计算，修改DRASTIC模型指标，将岩溶区地下水系统防污染性能评价采用地下水位埋深(D)、富水性(W)、土壤类型(S)、地形坡度(T)、包气带介质(I)、含水层补给量(R)6个因子评价潜水区地下水系统防污染能力，以 $DWSTIR$ 表示计算的指标值。

$$DWSTIR = 0.2D_i + 0.15W_i + 0.2S_i + 0.05T_i + 0.25I_i + 0.15R_i$$

$DWSTIR$ 值的范围为1.45~9.55，划分地下水防污染性能5级。

计算出济南岩溶水系统潜水分布区抗污染能力分区3个，$DWSTIR$ 值在4.6~9.15之间，不存在抗污染能力很好的Ⅰ级区和抗污染能力好的Ⅱ级区，可见由 $DWSTIR$ 指标法的评价结果与济南地区的水文地质条件相符。

由于城市向南部发展直接影响泉水补给和地下水环境质量，为避免开发建设影响泉水补给，根据计算结果与水文地质条件对比，严格实施"南控"，必须保护平安店—潘村—玉符河—丰齐—大杨庄—刘长山—英雄山—羊头峪—牛旺一线以南，炒米店—卧虎山—锦绣川—西营以北地区。建议修编"2020年济南市总体发展规划"时，必须充分考虑新城布局与地质环境保护之间的关系。"西进"规划宜在段店—大杨庄—大金庄—峨眉山—古城—朱庄以北。

3.7 玉清湖水库实施水资源转换

据调查，由于水库渗漏，不仅玉清湖水库的供水效率较低，而且造成库区外围沼泽化。因此，利用地下开采截取水库渗漏量，诱发水库补给增量，可以起到改良水质、防止耕地沼泽化和提高水库供水效率的三重作用。根据安全坡降下渗流计算，地下开采截取水库渗漏量大于10万 m^3/d。

3.8 进行优化配置，实施多水源供水

多水源供水原则是先用客水，后用当地水；先用地表水，后用地下水；先生活用水，后生产用水，实现当地地表水、客水(引黄、引江)与地下水联合调度，逐步建设水资源管理预警系统。

地下水地表水库联合调度分析

刘本华

(济南大学城市发展学院，山东，济南 250002)

1 概述

20 世纪 50 年代以来，随着西方工业社会的经济、人口的发展以及人们对生活环境要求的提高，人们认识到水资源是一种不可再生的资源或是短时期内很难再生的资源，因此如何最有效地利用现有水资源成为人们研究的重点，如 Bank(1953 年)指出，地下水、地表水联合运用会产生经济效益；Todd(1959 年)指出，地下水、地表水联合利用比单独使用更为可靠。60 年代，发达国家开始对地下水、地表水联合规划开发进行研究，如 Cashe 和 Lindeborg(1961 年)以两个农业用水区效益最大为目标，建立了线形规划模型来优化地下水、地表水的联合运用。Buras(1969 年)利用三状态变量动态规划方法，确定了由多个地下水库和多个地表水库所筑成的复杂系统的最优库容、抽水策略及地表水放水策略。

我国对地下水、地表水的联合调度的研究也有多年的历史，如翁文斌等对北京房山大石河流域地下水与地表水联合运用进行了研究，结果表明"联合利用比单独用地表水节省资金 $2000×10^4$ 元"；韩再生对秦皇岛石河流域地下水、地表水库联合开发利用的研究同样表明"联合开发比单独供水投资效益好，年平均增加可利用水量 $1300×10^4 m^3$"。刘梅冰等对福建省地下水与地表水优化配置的研究，通过分析福建省地表水和地下水的资源情况、开发利用现状及其存在的问题，提出了进一步开发地下水，并与地表水进行优化配置的可行性及若干对策；赵春锁通过研究南水北调供水区水资源优化配置问题，提出各种水资源优化配置方案以及不同水情、工情下的水资源调度原则，为河北省南水北调工程实施后的管理调度提供了科学依据；王维平等结合已开工的南水北调东线工程山东段和胶东调水工程，建立了山东省水资源优化配置模型，按照现状年、2010 年和 2030 年共 3 个水平年的 6 个方案进行了计算机模拟计算，探讨了山东省当地水、黄河水和长江水联合运用和优化配置，为新时期山东省水资源开发利用决策提供了科学依据。

2 地下水、地表水联合调度

2.1 地下水、地表水联合调度系统组成

地下水、地表水联合调度系统由地下供水系统、地表供水系统和用水系统三部分组成。地下供水系统主要包括地下水源地和相应的输配水系统；地表供水系统主要包括水源工程和输配水系统。

2.2 地下水、地表水联合调度模型中目标函数的选择

地下水、地表水联合调度的目标函数主要是在一定的约束条件下，使得供水或效益最优，常用的目标函数主要有：

(1)在一定投资规模或开发规模下，使得总供水量为最大；
(2)在一定供水需求条件下，使得总投资费用最低；
(3)在一定投资规模下，使得可用水资源量最大；
(4)在一定开发规模下，使得经济效益与生态效益最佳。

2.3 地下水、地表水联合调度模型的优化问题

地下水与地表水是两个具有较大差异的系统，组合在一起进行水资源的综合调度，具有一定的复杂性，这样的联合调度模型要考虑到地表水、地下水之间相互联系和相互制约的关系特点。在此我们仅就地表水与地下水存在直接水力联系的调度系统的优化问题进行一般性的论述。

2.3.1 不包含河流取水工程时的地下水、地表水联合调度问题

该系统是根据地下供水系统与地表供水系统的相对位置，采用顺序优化的方法，先建立上游地表供水调度模型，再建立下游地下供水调度模型，最后进行协调优化。

2.3.2 包含河流取水工程的地下水、地表水联合调度问题

河流取水工程无调蓄能力，主要依靠天然的地表径流，丰枯年份水量变化较大，因此包含河流取水工程的地下水、地表水联合调度遵循如下的原则：

(1)优先利用河川径流，在丰水季节，尽可能地利用地表水，减少弃水；

(2)在干旱季节河水供应不足时利用地下水来补充以满足需水量的要求。

2.3.3 对于建立地表水库的情况

遵循先利用地表水库中的水，枯水季节在满足一定要求供需平衡的条件下，大量开采地下水库中的水，为丰水期腾出库容，减少地表弃水的排放。

根据上述各自的特点，模型的求解首先是计算河流引水工程的取水量，与所需水量比较，确定不能满足需水的水量和时间，然后依照相对位置进行求解，最后优化协调，以达到水资源利用的优化目的。

3 莱钢总厂付家桥地下水库、地表水库联合调度实例

3.1 付家桥概况

山东省莱芜钢铁总厂厂区附近为丘陵地形，地表及河道坡降较大，大气降水形成地表径流沿河道排向下游。莱钢生产、生活用水均以开采地下水为主。目前，主要供水水源有付家桥、丈八丘、一铁、东泉及新城五个水源地。

随着莱钢总厂的发展扩大，现有的储量已不能满足工农业生产的需要，地下水超采，地下水位逐年下降，形成较大范围的降落漏斗，使得水资源供需矛盾日益突出，直接影响了总厂的改扩建及地方的长远规划。因此，需在付家桥一带修建地表水库，最大限度地拦蓄地表径流，强化对地下岩溶水库的补给，发挥本区地下岩溶水库的多年调节作用。选择付家桥、丈八丘、一铁水源地作为地下水、地表水调度计算的区域，本着"以补定采，采补平衡"的原则，增加均衡开采量，提高莱钢总厂及钢城区工业与城市供水能力，作为解决当前水资源供需矛盾的主要途径。

3.2 水文地质条件

研究区的东部以F1断层为界，西部以F2断层为界，北部以F3、F4断层为界，南部山区为奥陶系、寒武系地层出露区。地表分水岭与地下分水岭基本一致，大致沿奥陶系中统马家沟组第一段灰岩顶出露界限确定为计算区的南部边界。由此构成了一个封闭的单一水文地质单元，东、西、北部为阻水边界，南部作为第二类边界。

地下岩溶水库含水层岩性为马家沟组第二至第六段灰岩，总厚度达830.0 m，由勘探查明岩溶发育深度达200.0 m，埋深130.0 m以上裂隙岩溶发育良好且较为均匀，考虑到显效库容，取埋深130.0 m作为地下岩溶水库的底部边界(即水库的底板)。

因区内多为灰岩裸露，裂隙岩溶发育，地下水、地表水联系密切，据本区岩溶地下水流场分析，具有统一的自由水面，故可视为非均质各相同性平面二维潜水。

3.3 有限元数值模型

根据水文地质条件,可建立有限元数值模型(I)如下:

$$(\text{I})\begin{cases} \dfrac{\partial}{\partial x}\left[K(h-b)\dfrac{\partial h}{\partial x}\right]+\dfrac{\partial}{\partial y}\left[K(h-b)\dfrac{\partial h}{\partial y}\right]+w=\mu\dfrac{\partial h}{\partial t} & (x,y)\in \overline{G}, t>0 \\ h(x,y,t)|_{t=0}=h_0(x,y) & (x,y)\in \overline{G} \\ h(x,y,t)|_{T_1}=h_1(x,y,t) & (x,y)\in T_1, t>0 \\ \left[K(h-b)\dfrac{\partial h}{\partial x}\cos(n,x)\right]+\left[K(h-b)\dfrac{\partial h}{\partial y}\cos(n,y)\right]|_{T_2}=-q(x,y,t) \\ & (x,y)\in T_2, t>0 \end{cases}$$

式中:K 为渗透系数,m/d;μ 为给水度;$b(x,y)$ 为岩溶含水层底板高程,m;$h(x,y)$ 为在区域 $\overline{G}$ 上的初始水位,m;n 为法向向量;T_1、T_2 分别为第一类、第二类边界;$q(x,y,t)$ 为单宽流量;$\overline{G}$ 为地下岩溶水库的计算区域;$w=\varepsilon(x,y,t)-\sum_{i=1}^{\nu}Q_i\delta(x-x_i,y-y_i)+E(x,y,t)$,其中,$\varepsilon=\varepsilon_1+\varepsilon_2$,$\varepsilon_1$ 为除地表水库以外的地表水垂向补给强度,m/d;ε_2 为开采强度,m/d;$E(x,y,t)$ 为地表水库的入渗补给强度,m/d;Q_i 为第 i 个开采井的开采强度,m/d;ν 为井个数。

对于地表水库,建立如下的水均衡模型(II):

$$(\text{II})\quad \overline{\omega}=\overline{X}+\overline{Q}_{入}-\overline{Z}-\overline{Q}_{出}-\lambda$$

式中:$\overline{X}$ 为地表水库库区大气降水量,m³/d;$\overline{Q}_{入}$ 为地表水库上游入库水量,m³/d;$\overline{Z}$ 为地表水库库区水面蒸发量,m³/d;$\overline{Q}_{出}$ 为地表水库弃水量,m³/d;λ 为地表水库对地下岩溶水库的入渗量,m³/d;$\overline{\omega}$ 为侧向径流排出量,m³/d。

3.4 地表水库与地下水库的联合调度

根据地表水库对地下岩溶水库的入渗量与地表水库水位和地下岩溶水库水位间的相互关系,建立水量联合调度模型(III):

$$(\text{III})\quad \lambda(x,y,t)=K_z\dfrac{\tilde{h}(x,y,t)-h(x,y,t)}{\tilde{b}(x,y)-h(x,y,t)}\tilde{S}(x,y) \quad \tilde{S}(x,y)\in \overline{G}, t\geq 0$$

式中:K_z 为垂向渗透系数,m/d;$\tilde{h}(x,y,t)$ 为地表水库水位,m;$\tilde{b}(x,y)$ 为地表水库的库底高程,m;$\tilde{S}(x,y)$ 为地表水库库区面积,m²;$\overline{G}$ 为地表水库库区,m²。

通过模型(I)、(II)、(III)耦合在一起,对地下水、地表水联合调度进行优化配置分析,并对地下水开采前景进行预报分析。

计算区域面积 12.034 km²,据地下水流场特点进行三角剖分,计剖分成 299 个单元,179 个节点。

据野外抽、压水试验资料及实测岩溶地下水等水位线分析,考虑不同含水层组、岩性构造、埋藏条件、岩溶发育特征,进行水文地质参数分区,将参数 K、μ 划分为 14 个参数区,分别给出初值(见表1)。

表 1 计算区分区参数表

参数分区		I	II	III	IV	V	VI	VII
分区面积(km²)		0.85	1.67	1.05	1.70	1.04	0.28	0.840
参数	K(m/d)	18.5	6.50	6.450	4.400	0.082	10.048	0.580
	μ	0.017	0.014	0.082	0.022	0.012	0.007	0.000 48
参数分区		VIII	IX	X	XI	XII	XIII	XIV
分区面积(km²)		0.67	0.40	0.77	0.98	1.02	0.72	0.25
参数	K(m/d)	30.00	0.250	10.30	4.000	0.335	0.500	0.011
	μ	0.020 5	0.002	0.198	0.018	0.009 85	0.004 5	0.007 6

3.5 根据实际资料处理补、排项

3.5.1 降雨量及入渗系数

采用莱钢气象站提供的1988年降雨量577.1 mm。直接补给区以奥陶系中统碳酸盐岩为主，岩溶发育，平均入渗系数0.32，间接补给区以砂页岩及变质岩为主，入渗条件较差，平均入渗系数0.2。

3.5.2 地表入渗量

牟汶河上游中段流经计算区的中部，采用实测断面资料。

3.5.3 侧向补给量

根据水均衡计算及岩溶地下水等值线的水力梯度确定单宽流量初值，经计算机校正好给出。

3.5.4 开采量

采用各个水源地各井的实际开采量(仅限于模拟、验证阶段)，部分农业用水及生活用水以开采强度形式给出。

经过模拟、验证后，得到各区正确的参数后，即可在不同条件下进行各个方案的预报。

3.6 地表水、地下岩溶水联合调度均衡开采量计算

在模型模拟、验证的基础上，对天然条件下及修建地表水库后的均衡开采条件下的地下水位作出调算。

3.6.1 调算原则

(1)选择1988年9月1日统测岩溶水位为起调时刻水位。

(2)采用长系列真实年法，系列代表年限为1964年6月到1993年5月，系列长29年计348个时段，系列代表性合理。

(3)为保证莱钢及钢城区地方工业供水，取供水保证率 p=97%。

(4)地下岩溶水库最低控制调节水位埋深为60 m。

(5)枯水季(年)动用的地下水库的储存量，必须在整个系列调算中全部补上，即调算末刻水位必须恢复到起调时刻水位，满足"以补定采，以丰补欠，多年均衡开采"的原则。

3.6.2 方案的确定

一方案：天然条件下，以均衡量开采条件下的地下水位；

二方案：修建地表水库后兴利水位为261 m时均衡开采条件下的地下水位；

三方案：修建地表水库后兴利水位为263 m时均衡开采条件下的地下水位；

四方案：修建地表水库后兴利水位为265 m时均衡开采条件下的地下水位。

3.6.3 调节计算

通过有限元法调算分析，其调节计算成果如表2所示。

表2 付家桥地下水库、地表水库联合调度均衡开采量计算成果

计算方案	兴利水位(m)	总均衡开采量(万 m³/d)	增加的均衡开采量(万 m³/d)	均衡期弃水量(万 m³/d)	供水保证率(%)	地下岩溶水库调节库容(万 m³)	地表水库调节库容(万 m³)	总调节库容(万 m³)
一方案		2.182		4.871	97.8	563.52	0	563.52
二方案	261	3.772	1.590	3.000	88.2	1 253.04	226.91	1 478.95
三方案	263	4.512	2.330	2.218	97.2	1 447.41	388.40	1 838.81
四方案	265	4.854	2.671	1.852	98.7	1 779.15	585.68	2 364.83

通过联合调度计算结果表明：

天然状态(一方案)下，因未建地下水库，在供水保证率97.8%的情况下，均衡开采量只有218万 m³/d，地表水大部分变为弃水而排向下游，故要增加地下水开采量，供水保证率下降，否则将破坏采补平衡原则，造成地下水超采，地下水位持续下降，均衡状态遭受破坏，水资源逐渐枯竭，对环境产生一系列不良影响。

修建地下水库后，日均衡开采量有显著增加，比天然状态(一方案)分别增加 73% ~ 122%，效益十分显著。但由于受地下岩溶含水层导水性的限制，地表水不能短时间内全部补给地下水，均有不同程度的地表径流弃水。故此，需要在岩溶发育地段，增加高压渗井以加大地表水的入渗强度，以更大限度地利用地下岩溶水库库容的多年调节作用，进一步增加地下水的均衡开采量。

从水资源角度而言，选择第四方案最优，同时考虑到工程地质条件以及库区库容与淹没面积之比等因素，选择兴利水位 265 m 是最为理想的方案。

3.7 地表水、地下水联合调度调蓄能力分析

修建地表水库，一在时间上充分调节了丰枯周期水资源的变化，二在空间上充分利用地表水库、地下岩溶水库的调蓄库容进行综合调度，达到以补定采、以丰补歉，实现在供水保证率条件下最大限度地增加均衡开采量的目的，大幅度提高水资源利用率。

联合调度计算结果表明，影响地下水库调蓄能力的主要因素：一是地表水库库容，二是开采井的布局。

3.7.1 地下水库调蓄能力与地表水库库容有关

未建地表水库库容为 0，在均衡周期内，丰水季地下水位上升，如 1964 年 9 月末、1980 年 8 月末，地下水位上升至最高水位，其地下水库库容为 4596 万 m^3，枯水季(年)，特别是连续枯水年，地下水位下降至最低点，其地下库容为 4033 万 m^3，而此时最大调节库容仅 563 万 m^3，地下岩溶水库库容利用率极低，未能最大限度地发挥地下库容的调节能力。

修建地表水库后，水库拦洪蓄水，增加了入渗补给时间，为增加均衡开采量提供了补给保证。由表 2 可看出，随着地表水库库容的增加，各方案地下岩溶水库的调节库容也有明显的增加，从而提高了地下水库库容的利用率，把全年的水量重新分配，显示出地表水、地下岩溶水联合调度的巨大优势。

3.7.2 库区开采井布局影响水库调蓄能力的发挥

通过优化管理模型的分析，库区现有开采井主要集中于付家桥、一铁、针 1、丈八丘四个水源地，布局不合理，造成局部水位降幅过大，布局不合理性在于动用了多年储存的地下水，而这部分水有可能永远都补不回来，更由于岩溶入渗在时间上存在滞后性，一方面地表水大量流失，另一方面地下水库中无水，地下水位持续下降，影响了调蓄能力的最佳发挥，在充分考虑含水层岩性、渗透参数以及开采井影响半径等多种因素的基础上，建立优化管理模型，对开采井的布局做出合理调整，从而最大限度地发挥地下水、地表水联合调度所带来的优势。

4 水资源开发利用对策

4.1 加强地下水、地表水的统一规划和管理

解决流域或地区水资源短缺问题，要立足于流域内现有水资源量，对其进行优化配置，确定工农业发展规模，注重产业结构的调整和节水措施的落实；同时加强地下水管理，强化地下水取水许可管理，严格控制超采区内新增加取水项目的审批，建立地表水、地下水联合调度机制。

4.2 建设节水型社会，提高水的利用效率

积极发展节水型农业，是建立节水型社会的首要任务；城市生活用水再利用工作仍然存在巨大的潜力。同时提高工业用水效率和效益，以高新技术为依托，调整产业结构，发展耗水低、环境破坏小的产业，逐步杜绝高耗水行业的发展。

4.3 加强水环境保护措施，加大地表水污染的惩治力度

现阶段随着经济的发展，地表江河湖泊的污染程度日趋严重，地表水的下渗又逐渐造成地下水的污染，而对于我们这样一个水资源本来就很缺乏的国家，不仅是一种浪费，更是一种犯罪。因此，加强水污染的惩治力度，保护水环境是时代发展的需要。

4.4 建立完善地下水动态监测系统

随着经济的发展，各个地区各个城市都不同程度地存在地下水超采问题，由于检测力度不够，对地下水动态缺乏了解，同时又没有引起人们的足够重视，长此以往，造成了各个城市或地区地下水漏斗的不断扩大，不仅使得很多机井报废，同时还造成了很多的社会经济问题，为了更好地保护珍贵的水资源，需要建立完善的地下水动态监测系统，全面掌握地下水动态。

总之，鉴于当前水资源问题的严重形势，加强水资源的规划和管理，特别是地下水、地表水联合调度是至关重要的。

参 考 文 献

[1] 邓继昌，刘本华，贾连杰. 有限元法在北方岩溶地区中地下水资源评价中的应用[J]. 工程勘察，2000(2).

[2] 刘本华，邓继昌，谷照升. 王河地下水库水资源的合理开发利用研究[J]. 吉林农业大学学报，2004, 26(2).

[3] 杜文堂. 对地下水与地表水联合调度若干问题的探讨[J]. 工程勘察，2000(2).

[4] 刘本华. 傅家桥水库地表水、地下水联合调度计算[J]. 工程地质，1996, 3(46).

[5] 刘梅冰，陈兴伟. 福建省地下水与地表水优化配置初步研究[J]. 福建地理，2004(3).

[6] 于京要. 河北省南水北调供水区水资源优化配置研究[J]. 河北水利水电技术，2002(21).

[7] 张平，赵敏，郑垂勇. 南水北调东线受水区水资源优化配置模型[J]. 资源科学，2006(5).

[8] 王维平，杨金忠，何庆海，等. 区域水资源优化配置模型研究[J]. 长江科学院院报，2004(5).

[9] 张文鸽，黄强，管新建. 区域水资源优化配置模型及应用研究[J]. 西北农林科技大学学报(自然科学版)，2005(12).

[10] 于建海，温国敬. 优化配置合理利用水资源[J]. 水利科技与经济，2006(5).

[11] 赵惠，武宝志. 东辽河流域水资源合理配置对工业用水的影响分析[J]. 东北水利水电，2004(11).

[12] N.伯拉斯. 水资源科学分配[M]. 戴国瑞,等译.北京：水利电力出版社，1983.

济南市地下水保护供水保泉成本效益分析

李庆国

(济南大学城市发展学院,山东,济南 250002)

1 概述

为实现济南市地下水保护行动计划目标,济南市区应该在充分引黄河、长江水前提下,工业用水采用地表水,生活用水采用地下水,新城区分质供水,农业灌溉在节水的前提下采用地表水或中水。生态环境用水一律使用地表水。供水结构发生了变化,地下水减少,地表水增多,由于引黄、引库及地下水开采需要单方成本不同,尤其地下水开采成本最低,现在地表水比例增加,在经济上是否合理非常重要,这影响到各方利益及供水保泉计划的顺利实施。因此,需要对供水保泉的成本及效益进行分析。

2 济南市区主要水源

目前济南市供水系统为多水源供水格局。市区水源主要由引黄工程黄河水、地表水库蓄水、地下水 3 个主要部分组成。

引黄供水:鹊山水库、玉清湖水库通过黄河水厂、玉清湖水厂向市区供水。

地表蓄水工程:地表水库中只有卧虎山水库、锦绣川水库通过分水岭水厂、南郊水厂向市区供水。南郊水厂作为济南市南部两个主要水厂之一,其水源地卧虎山水库一直使用暗渠送水。分水岭水厂的水源来自锦绣川水库,输水大部分为明渠。

地下水:东郊、西郊、市区水厂开采地下水及市区企业自备井开采地下水。

3 济南市区各水源供水量调查统计

市区供水主要包括工业生产用水与城镇生活用水。济南市 2002~2004 年不同水源供水量统计见表 1。其中,地下水供水量中自备井部分供水量按 20 万 m^3/d,即以 7300 万 m^3/a 计。

表 1 济南市 2002~2004 年供水量统计

年份	引黄(万 m^3)	地表水库(万 m^3)	地下水(万 m^3)	合计(万 m^3)
2002	15 958.3	2 737.7	15 213	33 909
2003	14 008.1	4 132.9	15 430	33 571
2004	12 876.8	6 410.2	17 881	37 097

注:引黄水量已扣除水库蒸发、渗漏及输送损失。

4 济南市区不同水源供水成本计算分析

4.1 计算依据

规范:采用水利部《水利建设项目经济评价规范》(SL72—94)。

其中成本:年运行费包括工资及福利费,材料、燃料及动力费、维护费和其他费用等,可分项计算,也可按项目总成本扣除折旧费、摊销费和利息净支出计算。

流动资金包括维持项目正常运行所需的全部周转资金。

水利建设项目借款按年计息。建设期利息计入固定资产；正常运行期利息计入项目总成本费用；运行初期的利息可根据不同情况分别计入固定资产或项目总成本费用。

效益：水利建设项目的财务收入包括出售水利产品和提供服务的收入。年利润总额包括出售水利产品和提供服务所获得的年利润，按年财务收入扣除年总成本费用和年销售税金及附加计算。首先弥补上年度的亏损，再按有关规定缴纳所得税，而后再按财会制度进行分配。

4.2 各水源供水成本计算

供水成本计算中，供水管理费、管线费用共同分担，其他各水源供水成本中购水费用、动力费用、制造费用、还贷利息及其他费用根据不同供水水源单独计算。

根据市供水公司提供资料，2002年供水管理费、管线制造费共发生6 852.3万元，2003年及2004年变化不大，都采用此值计算。这样，由于除去自备井部分，2002年供水量为26 609万 m^3，则单方水由于管理费用增加成本0.25元。2003年单方成本增加0.26元。2004年单方成本增加0.22元。

其他单独费用计算如下。

4.2.1 引黄工程供水

引黄工程供水成本主要包括玉清湖水库、鹊山水库、水厂建设管理成本，以下对2002~2004年度引黄工程供水成本分别进行分析计算。

4.2.1.1 2002年费用计算

玉清湖水库、鹊山水库、玉清水厂：固定资产折旧6 442.22万元；水库建设借款利息5 947.71万元；动力费2 957.72万元；水资源费828.70万元；其他防汛、水处理费用619.63万元。

黄河水厂：生产成本1 284.9万元；制造费用995.8万元。

合计为：19 076.68万元，单方成本1.44元。

4.2.1.2 2003年费用计算

2003年费用中，固定资产折旧、还款利息仍按2002年值计算。动力费、水资源费及其他如水处理费按引水量进行分析计算。

玉清湖水库、鹊山水库、玉清水厂：固定资产折旧6 442.22万元；水库建设借款利息5 947.71万元；动力费2 596.28万元；水资源费727.43万元；其他防汛、水处理费用543.91万元。

缺黄河水厂资料，按2002年成本计算，则黄河水厂：生产成本1 284.9万元；制造费用995.8万元。

合计为：18 538.25万元，单方成本1.58元。

4.2.1.3 2004年费用计算

2004年费用中，固定资产折旧、还款利息仍按2002年值计算。动力费、水资源费及其他如水处理费按引水量进行分析计算。

玉清湖水库、鹊山水库、玉清水厂：固定资产折旧6 442.22万元；水库建设借款利息5 947.71万元；动力费2 386.58万元；水资源费668.68万元；其他防汛、水处理费用499.98万元。

缺黄河水厂资料，按2002年成本计算，则黄河水厂：生产成本1 284.9万元；制造费用995.8万元。

合计为：18 225.87万元，单方成本1.63元。

4.2.2 地表蓄水工程供水

济南市通过济南市供水公司南郊水厂、分水岭供水公司从卧虎山水库、锦绣川水库引水。以供水公司外购单价0.26元作为购水单方成本。

2002年，南郊水厂购水340.4万 m^3，生产成本(包括动力费、工资等)为626.9万元，制造费用(包括折旧、职工工资、修理费用等)为353.5万元。

其他从水库引水水厂主要为分水岭水厂。由于缺乏资料，成本分析通过南郊水厂进行类比计算。南郊水厂目前每日供水量47 000多 m^3，而分水岭水厂最大达到每日52 000多 m^3 的供水量。通过供水量类比二者规模并计算分水岭水厂2002年生产成本为770.0万元，制造费用为390.0万元。

因此，得到2002年这部分水量承担的生产成本为1 396.9万元，制造费用为743.5万元。2003年、2004年费用计算，对于生产成本，可以通过供水量变化比例进行计算；制造费用采用2002年值计算。2002~2004年从水库供水成本见表2。

表2 济南市地表蓄水工程供水成本

年份	引水量 (万 m^3)	单位购水 成本(元)	购水费用 (万元)	生产成本 (万元)	制造费用 (万元)	总成本 (万元)	单方成本 (元)
2002	2 737.7	0.26	711.8	1 396.9	743.5	2 852.2	1.29
2003	4 132.9	0.26	1 074.5	2 108.8	743.5	3 926.8	1.21
2004	6 410.2	0.26	1 666.6	3 270.8	743.5	5 680.9	1.10

4.2.3 地下水

济南市区由东郊水厂、西郊水厂、市区水厂开采地下水，加工成自来水。2002年这部分水量承担的费用为：生产成本，包括动力费、工资等2 851.5万元；制造费用，包括折旧、职工工资、修理费用等973.3万元。

2003年、2004年费用计算，对于生产成本，可以通过供水量变化比例进行计算；制造费用采用2002年值计算。

其他还有一些企业自备井，这部分开采的地下水量不通过市供水公司（不计管理费），平均年供水约为7 300万 m^3。根据水资源费用为1.80元/m^3，自备井单方生产成本大约0.7元，因此自备井单方成本按2.5元计算，则这部分水量平均每年成本约为18 250.0万元，见表3。

表3 济南市地下水供水成本

年份	生产成本(万元)	制造费用(万元)	总成本(万元)	单方成本(元)
2002	2 851.5	973.3	3 824.8	0.73
2003	2 929.7	973.3	3 903.0	0.74
2004	3 917.5	973.3	4 890.8	0.68

5 济南市区供水效益计算

供水效益按实际水价计算。根据济南市供水公司提供的资料，目前居民生活水价按2.35元/m^3计，经营服务水价按3.5元/m^3计，工业水价按2.4元/m^3计。

自来水的水处理损失及水厂自用水按35%计算。那么，水利用系数为65%，各年在供水技术没有大的改进的情况下，本次计算2002~2004年水利用系数取相同值。根据上述各行业的用水效益和各行业的用水结构比例，计算供水的综合效益。计算结果见表4。

表4 济南市综合供水效益计算表

年份	供水效益(万元)		综合效益 (万元)	水利用系数(%)
	工业用水	城镇生活用水		
2002	19 835.4	24 240.7	44 076.1	65
2003	21 804.1	19 964.6	41 768.7	65
2004	23 259.6	23 981.4	47 241.1	65

以上计算不包括自备井供水效益部分。对于自备井部分，供水单价按 2.55 元/m³ 计算，则自备井供水部分平均每年效益为 18 615.0 万元。

6 济南市区供水成本效益分析

6.1 2002～2004 年供水保泉成本效益分析

对 2002～2004 年成本效益进行分析。三年总成本为 146 428.19 万元，产生效益为 111 408.0 万元，加上自备井部分 55 845.0 万元，合计为 167 253.0 万元。

6.2 《济南市地下水保护行动计划》供水保泉成本效益分析

根据《济南市地下水保护行动计划》的供水保泉远期控制指标及保泉目标，在保泉的前提下，合理开采地下水，泉域开采量控制在 16 万 m³/d，外围地下水开采量为 36 万 m³/d，那么地下水供水规模为 42 万 m³/d。见表 5。

表 5 供水保泉远期控制目标下年供水量　　　　　　　　　　　（单位：万 m³）

地下水	引黄工程	引江	引库	泉水先观后用
15 330	29 200	14 600	3 650	3 650

在保证大明湖和护城河等生态环境用水的前提下，积极利用泉水资源，建设泉水先观后用工程，泉水供水规模为 10 万 m³/d，主要用于生活用水。泉水先观后用工程计划投资 15 500 万元，根据《水利工程固定资产分类折旧年限的规定》，折旧年限取为 50 年(年折旧率 2%)，则运行期间年折旧费用为 310.0 万元，年运行费取总投资的 2.5%，则为 387.5 万元。相对于 2002 年，这部分是增加的费用。

充分利用当地地表水资源，卧虎山及锦绣川两水库联合向城市供水 10 万 m³/d，主要用于生活用水和景观用水。以黄河水和南水北调东线工程水为水源，利用鹊山、玉清湖和东湖等三大调蓄水库向城市供水 120 万 m³/d，其中鹊山水库、玉清湖水库和引江工程东湖水库分别供水 40 万 m³/d。

对引江供水部分成本费用单独计算。引江供水口门价格为 0.8 元/m³，水资源费根据济南市标准取 0.35 元/m³。水处理费用参考黄河水厂计算，2002 年黄河水厂生产成本 1 284.9 万元，制造费用 995.8 万元，供水 6 633 万 m³，则单方处理费用为 0.34 元。管理费取 0.25 元/m³。因此，2002 年水平引江供水单方成本为 1.74 元。

单方成本除引江部分单独计算外，其他按 2002 年水平计算；水价采用 2002 年标准。进行供水保泉远期控制目标下年供水总成本为 87 093.0 万元，可实现供水总效益 110 107.7 万元。

7 结论

按《济南市地下水保护行动计划》远期供水计划，水价如采用 2002 年标准，进行供水保泉远期控制目标下年供水总成本为 87 093.0 万元，可实现供水总效益 110 107.7 万元。可见，其在经济上是可行的。地下水保护行动计划的实施必须有经济投入作保障，建议将《济南市地下水保护行动计划》列入济南市社会与经济发展中长期规划，加大对地下水资源保护的投入。

参 考 文 献

[1] 济南市地下水保护行动计划[R]. 济南市水利局, 2006.
[2] 山东省水资源调查评价汇报提纲[R]. 山东省水文水资源勘测局, 2003.
[3] 山东省济南市保泉供水勘探报告[R]. 山东省地矿局 801 水文队, 1988.

[4] 济南泉域地下水补给区保护对策研究[R]. 山东省水利科学研究院, 2002.
[5] 济南市岩溶地下水开发利用与泉群保护研究[R]. 山东省水利科学研究院, 2003.
[6] 济南保泉综合技术研究[R]. 济南市水利局, 2005.
[7] 河南省地下水保护行动计划[R]. 河南省水利厅, 2003.
[8] 济南市统计年鉴[R]. 济南市统计局, 2003.
[9] 济南市水长期供水计划[R]. 济南市水利局, 1994.

济南市地下水补给水源涵养能力分析

周保华[1,2] 李大秋[3]

(1 中国矿业大学(北京校区)资源与安全工程学院，北京 100083
2 济南大学城市发展学院，山东，济南 250002
3 济南市环境保护科研所，山东，济南 250014)

济南的地下水属于深层岩溶地下水，水温稳定，常年保持在 17~18 ℃，微量元素含量丰富，硬度低、矿化度低，是优质的饮用水源[1]。南部山区经过长期的溶蚀和多次构造运动，岩溶地貌发育，形成由地下溶洞、溶沟和溶隙等构成的地下输水网络，大气降水和地表径流在南部山区的灰岩出露或裂隙岩溶发育区渗入地下补给地下水，并呈扇形网络状顺地势向北潜流，至城区遇不透水侵入岩体阻挡，在巨大承压作用下，地下水潜流由水平运动变成垂直运动，穿过裂隙和第四系覆盖层，以泉群的形式喷涌到地面，形成了闻名于世的"泉城"。然而，多年来的不合理开发利用，致使泉水自1972年春季开始出现断流现象，究其原因，关键在于济南市南部山区地下水补给水源涵养能力大大降低。为增加地下水的补给量，进行南部山区水源涵养能力分析研究具有重要意义。

1 济南市地下水补给区概况

作为济南地下水补给区的南部山区以寒武系凤山组与中下奥陶系厚层灰岩为主，总厚度 800~1 000 m，岩溶裂隙发育，地下连通性好，水力联系密切。受构造、地形和含水层埋藏条件影响，其富水性差别较大。钼酸铵和孢子真菌示踪试验表明[2]，地下水呈扇形交叉网络状结构展开，总体由南向北潜流，水力坡度在南部山区为 1.5%～2.5%，由南向北且由东至西，水力坡度呈逐步减缓的趋势。济南泉域地质构造决定了该区域大气降水—地表水—地下水—泉水"四水"转化直接而且迅速，形成了地表水在南部山区补给区涵养渗漏补充、地下水快速运移、承压区排泄的"补给、径流、排泄"方式。

区域的水源涵养能力主要与降水、蒸发、径流、植被、土壤、地质、坡度等因素有关。区域内 3 个雨量站多年平均降水量为 710 mm，北沙河和玉符河两条河流流量控制站的多年平均径流深约 178 mm，该流域的多年平均蒸发量为 2 060 mm[3]。区域河流基本上为季节性河流，洪水历时较短约 90 d，并且由于水库的层层拦截，河水的基流不大，甚至常常干涸。南部山区靠近市区的为直接补给区，距市区较远的是间接补给区，其地质与坡度因素差别较大。直接补给区岩石透水性和水蚀性较强，坡度相对较缓，形成的地面径流小，入渗系数为 0.33～0.45[4,5]，水源能较多地补给地下水；而间接补给区岩石的透水性和水蚀性较弱，坡度相对较大，地表径流多流向下游，入渗系数在 0.20～0.30 之间。除以上因素外，植被与土壤是影响地下水水源涵养能力最关键的要素[6]。

南部山区地处鲁中山地的北缘，以低山丘陵为主，地形起伏度大，植被覆盖率低，水土流失严重，是全市水土流失最严重的地区，属生态脆弱区[7]。近30年来的城区不断南扩，地面硬化面积加大，南部山区地下水补给区的无序开发和地下水盲目、过度开采等，致使地下水位急剧下降，泉水断流现象日渐加重。据调查，目前矿区使用土地面积达 112.35 hm^2，固体废弃物的堆积占地面积 1.48 km^2；在本区矿山开发的鼎盛时期，共有矿山上百座，占地达到几百亩；水土资源大量流失，流失面积达 863 km^2，占研究区面积的 77%(济南市国土资源局 2005 年调查资料)。济南南

2 地下水补给区不同植被生态系统水源涵养能力分析

区域内最大的绿地斑块—龙洞(林木覆盖率达 75%以上),综合体现了林地的水源涵养功能,因而位于龙洞的老君井流水常年不断。然而,不同植被覆盖的地表涵养水源的能力是有差别的。

2.1 区域内不同植被分布情况

在 2000 年 6 月卫星图像解译数据基础上,用 TM 与 SPOT 影像相融合(分辨率 10 m),并辅以该年度Ⅱ类林地资源实地调查资料,确定了济南地下水补给区的不同植被的分布状况。结果表明,农田是区域最主要的生态系统,其面积占总面积的 37.6%;其次为林地和草地生态系统,分别占 29.5%和 15.9%。另外,城乡工矿居民用地、裸岩地和水域生态系统分别占 11.8%、3.9%和 1.4%。

解析数据表明:在林草生态系统中,疏林地占区域总绿地面积最大,为 21.8%;其次是有林地,占区域总绿地面积的 19.5%;而中和低覆盖草地(覆盖率 0.2～0.5 和<0.2)则占区域总绿地面积 25.1%。可见,区域内绿地功能较差的疏林地和中、低覆盖草地所占面积较大,对地下水补给水源的涵养不利。

2.2 区域内植被生态系统涵养水源的能力分析

不同植被的截留渗蓄能力相差较大,根据 14 块样地的试验观测资料与区域的相关研究,确定不同植被类型的截留渗蓄系数,应用水量平衡法来计算评价不同植被的水源涵养能力,即:

$$W_1=(R-E_i)A_i a = RA_i \theta_i a$$

式中:W_1 为涵养水源量,m^3/a;R 为平均降水量,mm/a;E_i 为第 i 类群落的平均蒸散量,mm/a;A_i 为第 i 类群落区域的面积,hm^2;θ_i 为第 i 类群落区的径流系数;a 为换算系数。

经过计算和 GIS 叠加识别,涵养水源的能力为:林地>灌丛>草地>农田>裸岩,复层混交林>单层纯林,阔叶林>针叶林。区域中的有林地以侧柏林、刺槐林和松林为主,所占的比例分别为 67.1%、15.4%和 13.2%,多为郁闭度大于 0.3 的次生林和人工林,所有有林地斑块的样地平均持水能力为 530 t/(hm²·a);灌木林地多为郁闭度大于 0.4 的较好的次生林,平均持水能力为 490 t/(hm²·a);疏林地为郁闭度在 0.1～0.3 之间的疏幼林地,水源涵养能力有所降低,平均持水能力为 469 t/(hm²·a)。南部山区单纯的针叶林较多,约占林地的 80%,涵养水源的能力相对较差。农田季节性强,生长初期和收获期其水源涵养能力低,甚至低于草地;而农田区的片段林网群落或窄林带、林网、小片林地、林粮间作和孤立木,树种单一,又伴生农作物,层次单一,透光度大,缺乏凋落物层,水分蒸发快,保水性较差。以苹果园最多、桃与板栗次之的果园,林下枯落物层和乔木层的覆盖率与其他林地系统相比均较低,涵养水源的能力相对其他林地较差;而典型成片的针阔混交林区,土壤多为瘠薄的山地,以乔木为建群树种,伴生植物多,层次复杂,林内辐射量较低,风速小,凋落物层厚,保水性好,水源涵养能力最强。

3 地下水补给区土壤水源涵养能力

土壤持水与渗水是涵养水源的主体,一般约占涵养水源总量的 80%[9,10]。所以,根据第二次全国土壤普查和卫片解译数据,对区域各类土壤的含水状况和重点保护的土壤斑块进行了分析和识别。

土壤水是济南"四水"转换的一个重要环节,植被从土壤中吸收水分用于生长和蒸散,土壤的重力水又可直接补充地下水,根据区域内土壤的类型、分布和物理特性计算水源涵养能力(见表 1),其中每年形成的土壤重力水达 $1.513×10^7 m^3$,土壤凋萎水总量 $1.309×10^7 m^3$,毛管水总量 $3.938×10^7 m^3$。区域内土层厚度最大和含水能力最强的土类是普通褐土中均质中壤($Ⅱ_{a4}$)、黏层中壤($Ⅱ_{a6}$)、淋溶褐土均质中壤($Ⅱ_{c14/1}$)、黄土发育的石灰性褐土($Ⅱ_{b3}$)及山前平原中下部的潮褐土或地势低洼的河谷地带潮褐土(II_{d1})等；而通气孔隙最高、渗水能力最强的土类为棕壤性土(I_{e1})、褐土性土(II_{e1})、棕壤类残坡积中层砂壤($I_{a12/4}$)和均质轻壤($I_{a13/1}$)、褐土坡洪积物中层轻壤($II_{b13/4}$)和中层砂壤($II_{a12/4}$)。超过土壤田间持水量的水会继续下渗而成为地下水，沿各级地下疏水网路汇入济南泉域，因此土壤田间持水量与土壤重力水就像"土壤水库"。由大涧沟地区地下水观测井的试验得知，夏季第一场雨或大到暴雨时，井总是先涨混水、后发清水，也说明了这里有深厚的土壤含水层，降水通过土壤就地蓄渗也是增加地下水最经济的措施[11]。

表1 土壤涵养水源量评价结果表

土壤类别	II_{a4}、II_{b3}、II_{c1}、II_{c6}、II_{d1}	I_{a2}、II_{b1}	I_{a2}、II_{a1}	I_{a1}、II_{a1}、II_{e1}、II_{e3}
土层厚度(m)	>2	1.5~2	1~1.5	<1
斑块数(块)	24	14	94	154
面积(km^2)	124.3	28.3	504.9	821.9
田间持水量($\times 10^4 m^3$)	8 101.9	1 264.0	20 442.4	5 792.5
有效水含量($\times 10^4 m^3$)	5 349.6	1 110.5	13 044.4	2 826.4
重力水总含量($\times 10^4 m^3$)	2 196.4	380.0	6 658.9	3 097.6
单位面积重力水含量($\times 10^4 m^3/(km^2 \cdot a)$)	17.67	13.43	13.19	3.77

4 地下水补给水源涵养能力分布状况

植被的根系与须根串联缠绕分割土体，形成富有团粒结构的土壤，在强化降水入渗性能、增加土壤水容量、消除超渗径流等方面来保证降水顺利下渗并不断形成地下径流流入各级水网系统。本文运用地理信息系统技术，将南部山区的不同植被覆盖的生态系统现状类型、土壤类型、坡度和地质状况等进行叠加分析后，得到区域综合系统水源涵养能力分布图，其中水源涵养能力最强，需要保护的区域面积为 86.97 km^2，占总面积的 5.75%，主要分布在直接补给区的龙洞、千佛山一带与间接补给区的药乡、凤凰岭等林地区域和冲洪积平原地带；恢复区面积 421.49 km^2，占区域面积的 27.86%，具有较强的水源涵养能力，多位于地下水补给区生态环境现状较好的区域，但仍需通过采取各种措施，提高其水源涵养能力，使其恢复为保护区；而水源涵养一般的区域面积 1 004.54 km^2，占区域总面积的 66.39%，所占面积之大与水源涵养生态功能保护区的地位是不相称的，因而需要通过植被与土壤的改良，大力提高一般区的水源涵养能力。

5 结论和讨论

(1)研究区域内涵养水源的能力差异情况是：林地>灌丛>草地>农田>裸岩，复层混交林>单层纯林，阔叶林>针叶林。而地下水补给区单纯的针叶林较多，约占林地的 80%，因此涵养水源的能力相对较差。

(2)研究区域内土层厚度最大和含水能力最强的土类是普通褐土中均质中壤(II_{a4})、黏层中壤(II_{a6})、淋溶褐土均质中壤($II_{c14/1}$)、黄土发育的石灰性褐土(II_{b3})及山前平原中下部的潮褐土或地势低洼的河谷地带潮褐土(II_{d1})等。

(3)对不同植被、土壤和地质条件组合的综合水源涵养能力分析，发现研究区域内水源涵养能力最强的仅占区域总面积的 5.75%，而水源涵养能力一般的占区域总面积的 66.39%，所占面积之大与水源涵养生态功能保护区的地位是不相称的，同时对地下水补给量的提高是不利的。

综合考虑上述济南市地下水补给区水源涵养能力的差异情况，应及时采取措施，加大南部山区的水源涵养林的建设和管理；保护好水源涵养能力最强土壤地带，改良一般土壤和植被覆盖区的条件；合理开发，减少矿山开发带来的不利影响；建立生态功能保护区，提高地下水补给区水源涵养能力，以增加地下水的补给量。

参 考 文 献

[1] 任宝祯. 济南名泉说略[M]. 济南：黄河出版社, 2002.

[2] 济南保泉课题组. 济南泉水来源试验研究[R]. 1996.

[3] 张杰, 茅樵, 宋玉琴. 济南市玉符河回灌补源保泉研究. 水利水电科技进展[J]. 2002, 22(3): 19-20.

[4] 何师意, 冉景丞, 袁道先, 等. 不同岩溶环境系统的水文和生态效应研究[J]. 地球学报, 2001, 22(3): 265-270.

[5] 李大秋, 高焰, 王志国, 等. 济南泉域岩溶地下水水质变化分析[J]. 中国岩溶, 2002, 21(3): 202-205.

[6] 吴长文, 王礼先. 林地坡面的水动力学特性及其阻延地表径流的研究[J]. 水土保持学报, 1995, 9(2): 32-39.

[7] 王伟, 郑新奇. 济南南部山地开发对构建生态城的影响探析[J]. 国土与自然资源研究, 2003(2): 13-14.

[8] 李玉顺, 赵艳, 柴永昌. 济南市历城区矿山地质环境问题及其恢复治理[J]. 山东国土资源, 2004, 20(1): 51-53.

[9] 欧阳志云, 王如松, 赵景柱. 生态系统服务功能及其生态经济价值评价[J]. 应用生态学报, 1999, 10(5): 635-640.

[10] 袁道先, 蔡桂鸿. 岩溶环境学[M]. 重庆：重庆出版社, 1988.

[11] 黄春海. 地下水开发研究[M]. 济南:山东大学出版社, 1988.

浅议海侵区农民用水者协会在水生态修复中的作用

曲士松　王维平　孙小滨

(济南大学城市发展学院，山东，济南 250002)

1 概况

1.1 滨海地区海水入侵产生的原因和危害

莱州湾海咸水入侵区海水入侵面积 1 000 km²。气象、水文、地质、地形、风暴潮等自然因素，是海水入侵发生和发展的基础条件，过量开采地下水等人为不合理的经济活动是海水入侵发生和发展的决定因素。农业灌溉大量超采地下水导致海水入侵并对当地经济社会和水生态环境带来严重影响。由于长期山丘区大量拦蓄地表水，减少了入海径流量和地下水补给量，而平原井灌区农业生产快速发展，特别是自 1976 年后，机井数量增加，地下水开采量增大，当遭遇到 80 年代初期的连续干旱年，地下水采补不平衡，导致严重的海水入侵并不断发展。据不完全统计，全区已有 4.48 万 hm² 耕地处在侵染区，7 446 眼机井报废，3.33 万 hm² 耕地丧失了灌溉能力，0.33 万多 hm² 耕地产生了次生盐碱化，农业产量大幅度降低。海咸水侵染灾害使工业设备严重锈蚀，产品质量下降，经济效益降低，部分企业面临转产或搬迁的困境。地下水侵染后，人畜吃水发生了严重困难，造成本区有 39 万人吃水困难，群众不得不长期饮用劣质水，危害人民身体健康。

1.2 滨海平原地区经济快速发展与地下水脆弱性的矛盾

滨海平原地区经济发达，人口集中，外贸活跃，水资源供需矛盾尖锐。同时农业灌溉用水量大，以开采地下水为主，加之在山丘区修建了大量地表水拦蓄工程，减少了平原区河流补给地下水量和山丘区与平原区的侧向径流量。平原区地下水环境十分脆弱，遇到干旱年或连续干旱年，地下水超采，若地下水位低于海平面，则导致海水入侵，一旦发生海水入侵，50 年甚至 100 年都无法恢复。

1.3 莱州湾海咸水入侵涉及到农民的生存与发展

莱州湾海岸带以莱州市虎头崖为界，以西为泥质海岸，以东为砂质海岸。因此，莱州湾海水入侵一种是海水入侵，另一种是古咸水入侵。海咸水入侵是由于水开发利用程度在枯水年份已超过当地地下水承载能力，主要是农业灌溉用水量造成的采补不平衡，说明水资源已不能支撑社会经济的发展，是一个社会经济问题，涉及到几十万农民的生存和发展。

控制和治理海水入侵，领导者和科技人员必须同农民结合，将研究成果转化为生产力，解决海水入侵区的农业问题，推动农业发展。目前，我国水法规定从江河、湖泊、河流和地下取水均要缴纳水资源费，但至今农业取用地下水不收水资源费，滨海平原区工业和生活取水易管理，而农业灌溉取水处于无序管理状态，枯水年、枯水灌溉季节各家各户为了满足作物灌溉需水超采地下水，缺乏节水和保护生态环境的意识。不解决农民用水管理、生存及发展的问题，仅从工程技术上去解决，海水入侵很难得到控制。

2 农民参与水生态环境修复管理是防止海水入侵的有效措施

2.1 农民参与灌溉管理是改革的必然趋势

农民参与灌溉管理和成立农民用水者协会是当今国际上倡导的水利管理体制改革的重要内

容。农民参与和灌溉管理权的转让其含义就是吸收农民参与灌溉管理工作，把灌溉管理权下放给农民用水者组织。

世界各国几乎普遍认为现有许多灌区的用水管理不善，财政收支难以平衡，在很大程度上与管理体制有关。灌溉农业与农民的切身利益关系十分密切，不能一切都由政府专业机构包揽，而应充分调动农民用水户自身的积极性，吸收农民参与灌溉管理。这样不仅可以节约用水，而且可以减少国家补助，减轻国家财政负担。因此，近些年来，许多国际组织，如世界银行、国际粮农组织、国际灌排委员会都十分重视吸收农民参与管理，组织用水者协会，改革灌溉管理机制，并作为一个方向来倡导。特别是由世界银行、国际粮食组织和世界粮食计划署投资兴建的灌溉工程，都强调必须吸收农民参与管理和组织用水者协会。为此，近十多年来，世界各国特别是亚非发展中国家，在新灌区的建设和旧灌区的改造中，把下放管理权、组织农民用水者协会作为一项重要内容，并且取得了丰富经验[1]。当前，约有43个发展中国家正在进行将灌溉管理权移交给农民的改革，而对大多数发达国家而言，灌溉系统的管理早就是农民用水者自己的责任。

所有这些国家进行的管理移交过程中所采用的主要是农民参与式灌溉管理，简称PIM。当然，PIM的正常运行需要地方政府的大力支持[2]。在中国，基于PIM这一概念的农民用水者协会对灌区进行管理的改革得到了中国水利部的大力支持[3]。

2.2 农民用水者协会是农民自我管理和防治海咸水入侵的有效途径

为了加强农村水利工程管理，实行测水、量水、水费征收和控制地下水开采制度，建立农民用水者协会既规范了用水管理，又有利于生态恢复和保护。在海水入侵区建立农民供水者协会的目的不仅是解决工程运行费、维修费和折旧费问题，达到成本水价，更重要的是将农民组织起来，共同参与，通过有效措施，保护地下水不超采，就是保护自己子子孙孙生存的耕地不受破坏。

3 海水入侵区建立农民用水者协会的指导思想、内涵、内容、特点

3.1 指导思想

坚持以人为本，树立全面、协调、可持续的发展观，以保护海水入侵区水生态环境为目标，结合多余洪水利用地下水回灌工程和种植业结构调整，统筹规划，突出重点，建立海水入侵区农民用水者协会和示范区，改善农民的生存环境、生活和生产条件，促进农村脱贫致富。

3.2 农民用水者协会的内涵

农民用水者协会是由灌溉供水区、饮水供水区的受益农民自愿参加的群众性用水管理组织。经当地民政部门登记注册后，具有独立法人资格，实行独立核算，自负盈亏，实现经济自立。农民用水者协会负责所辖灌溉系统及供水系统的管理和运行，保证灌溉工程、供水工程资产的保质和增值。

协会由用水小组和用水协会会员组成。用水小组：是将供水受益区内的用水户按地形条件、水文边界、街道分布情况，同时兼顾村、村民小组边界划分的若干个用水组。通过用水小组大会选举的每个用水组的用水者代表即是该组的管水员，负责执行用水者代表大会作出的决定，管理本组的灌溉设施、供水设施和用水工作，向本组用水户收取灌溉水费和饮水水费，并负责本组的其他日常工作。用水协会会员：是指在协会范围内的受益农户户主，由用水户自愿申请加入，并经用水者协会批准。其权利和义务主要是：参加本组大会，有灌溉用水、人畜用水的权利和按时缴纳水费的义务，并根据要求参加保护、维护协会管理范围内的管排设施、供水设施和地下水采补平衡即不超采。

3.3 海水入侵区农民用水者协会的特点

农民用水者协会是具有法人地位的社会团体组织，在民政部门注册登记，取得合法地位，依照国家的法律、法规及协会章程运作，并对管理设施、财产和灌溉服务、人畜吃水供水服务负法律责任。它不是以营利为目的的服务性经济实体。它通常按行政村或村民小组组建，不受当地行

政机构的干预。农民用水者协会是农村从事灌溉服务、生态保护或人畜用水服务的民主组织,用水户按自己的意愿民主选举协会的组织机构,内部事务通过会员代表大会以民主协商的方式确定,活动受到用水户的广泛监督。协会受政府业务主管部门的服务监督和指导。海水入侵区农民用水者协会的作用就是民主生态恢复管理,让农民在基层对生态环境管理负有责任。

3.4 海水入侵区农民用水者协会的内容

农民用水者协会必须建立在广泛宣传的基础上,要对协会人员、用水者代表和广大用水户做大量的宣传。按自愿、公开、民主、因地制宜、依法、经济自立、水文边界与行政边界相结合的原则。

海水入侵井灌区一般以一眼机井为基本单元组建用水小组,由农民推选用水者代表;以一个电力变压器控制面积为单元,成立用水者协会。也可以自然村或行政村成立用水者协会,通过民主选举推荐用水者协会主席和执委会成员。即用水者协会的组织模式为"用水者协会+用水小组+用水户"。

用水者协会主要是对计划用水进行监督,协助政府加强地下水资源管理,指导用水小组开展节水灌溉和技术服务。各用水组为独立核算的经济组织。规划在海水入侵区成立的单个用水者协会,大田面积一般控制为 133~333 hm^2,以便管理。

自 1976 年后,莱州市滨海平原机井密度大幅度增加,农业产量大幅度提高,地下水开采量也同样大幅度增加,结果造成海水入侵这种灾难性后果。因此,在海水入侵区利用协会的力量,公平合理地封闭部分机井,示范推广农业灌溉节水技术包括农艺节水措施,调整农业种植结构,选用耐旱耐盐作物品种等,并通过协会,将农民组织起来,配合地下水回灌补源工程的实施。

3.5 政府部门的作用

政府提供资金用于农民用水者协会的开办、技术培训、宣传、建观测井、购买和安装测水量水设备,为农民用水者协会选择和配备适当的监测仪器和设备,例如水位计、电导仪,定期监测和发布地下水水位、水质信息。

政府资助进行与农民用水者协会相关的技术支持的研究、技术培训并配备有关仪器设备。由于海水入侵区农民用水者协会的建立不仅仅是用水管理,更重要的是如何不使地下水在枯水年或连枯年或枯水季节不超采,达到保护生态环境的目的。通过研究划定海水入侵区地下水开采的分区,例如禁采区、限采区和可采区;在限采区和可开采区,确定地下水开采的控制指标:最佳水位、警戒水位和破坏水位等。

由于海侵区水资源短缺,同时灌溉机井工程自建、自管、自用,且规模较小,政府部门对此应进行宏观管理,例如新建机井必须经过批准。另外,建立一种长效的灌溉农业旱灾补偿机制,使农民在特枯年既不超采地下水造成新的海水入侵又有饭吃。

4 结论及讨论

农民参与海水入侵区地下水灌溉管理,建立农民用水者协会,是防止和修复海水入侵引发的生态环境问题的有效措施,适合目前农民人均灌溉面积小、机井数量多和分散,仅靠政府很难管理的特点,符合国内外用水户参与灌溉管理的改革趋势。其目的是生态环境保护,禁止超采地下水。

讨论:

(1)农民用水者协会将农民自身的利益与区域的生态环境保护的责任结合起来,是海水入侵区水环境保护的方向。目前中国的农业生产是以家庭为最小生产单位,人均耕地 1/15 hm^2 左右,土地经营规模较小,实行家庭联产承包责任制,村级行政管理为基础的小农生产方式。农民追求每家每户经济效益最大,他们重点关心自己收成的好坏,能否在作物干旱的时候灌溉,抽取地下水的电费省一些,而对能否节水,保护大家的共同家园不受破坏关心得少,但是一旦生态环境被破坏，他们希望政府或集体给解决问题。国家规定农民土地承包 30 年不变，村级政府具有分配、调整土地的权利。随着取消农业税，农民将享受更多的利益，有利于农民用水者协会的建立。海水入侵问题涉及千家万户的农民，是一个面上的事，也是随降雨量多少不断变化。仅靠工程措施，很难解决好这些问题。将分散的农民通过农民用水者协会组织起来，在外界一定的技术指导和援助下，自我管理，而不是完全依靠村级政府，当然完全离开了村级政府，要想搞好水管理也不行。

(2)如何解决农民用水者协会的运行费是保证正常工作的关键。由于目前机井都是农民自建自管，通过收水费来提取运行费有一定困难，因此按每家每户的灌溉面积或按实际灌溉用水量来提取。

(3)由于农民用水者协会是农民自愿参加的社会团体组织，能否承担保护自己家园又能发展生产的责任，只有先试点，不断完善，取得经验后，再推广。

参 考 文 献

[1] 许志方. 农民参与管理和小型水利体制改革[J]. 中国农村水利水电, 2002 (6): 8-10.
[2] 理查德·瑞丁格. 中国参与式灌溉管理改革：自主管理灌排区[J]. 中国农村水利水电, 2002(6): 7-9.
[3] 李代鑫. 中国灌溉管理与用水户参与灌溉管理[J]. 中国农村水利水电, 2002(5): 23-26.

区域农业水资源联合调控技术研究

徐征和[1]　贠汝安[2]

(1 济南大学城市发展学院，济南 250002　　2 山东大学，济南 250100)

该研究设置在地处胶东半岛沿海地区的威海市环翠区，农业灌溉水源仅为小型水库、拦河坝、大口井、浅机井等分散的多种类型小水源，水资源严重缺乏，研究该区域向农业供水的分散水源之间的水资源如何优化配置，各水源之间水量如何调配，是实现该区域水资源高效利用的关键技术。

1 调控区域的基本情况

示范区面积 666.67 hm^2，有多种灌溉水源，机井出水能力 20~30 m^3/h，静水位埋深 2.5 m，动水位埋深 10~25 m；大口井出水能力 50~150 m^3/h；方塘出水能力 50~250 m^3/h。示范区现有小(2)型水库 1 座，兴利库容 15 万 m^3，塘坝 1 座，兴利库容 2 万 m^3，拦河橡胶坝 2 座，方塘大口井 12 处，机井 26 眼，扬水站 1 座，示范区多年平均总供水量为 640.5 万 m^3。距海最近距离不到 2 km，地下水的过度开采已引起地下水位下降，海水入侵。干旱季节地表水拦蓄工程基本干涸无水，农民只能靠过量开采地下水维持灌溉。规划将区域拦蓄的地表水与示范区内的小水库、大口井、方塘与机井通过地下管网联合调度，用于农田灌溉，平时向机井、方塘回渗补源，形成联合供水的良性循环。

2 区域农业水资源优化配置技术

2.1 区域水资源优化配置的总体思路

该系统是由自然系统和人工系统组成的综合系统，自然系统可概化为地表水系统和地下水系统，人工系统可概化为供水系统和需水系统，二者之间通过人类活动来联系。

水资源分配中，先确定生活用水和工业用水，再分配作物灌溉。本区域水资源优化配置研究主要是针对可用于农业灌溉的地表水、地下水水量在各种作物间的分配及在各作物不同生育阶段的水量分配问题。根据模型结构及项目区特点，确定模型目标函数、决策变量与约束条件，在满足约束条件的基础上确定作物种植布局预方案，然后根据现状年不同作物的灌溉需水定额，推求该方案下的作物灌溉需水总量，通过与现状年可供水资源量的比较，确定是否满足水资源供需平衡和模型精度的要求，否则，调整作物的种植布局方案，重新试算，直至满足精度 Ep 要求，最终确定项目区最优的作物种植布局方案，并对可供水资源量在各种作物之间进行优化分配。

2.2 目标函数

(1)经济效益目标。将区域经济效益之和最大，即区域净效益最大作为目标函数。

(2)粮食目标。粮食产量影响到政治稳定、经济发展、社会保障等众多政策性问题不容忽视。

2.3 水资源优化配置模型

不同作物之间的配水属于总系统的优化，而作物各生育阶段之间的配水则属于每个作物子系统的优化，将作物子系统作为第一层，总系统作为第二层，通过分配给每种作物的供水量将两层联系起来，则成为一个具有两层谱系结构的大系统，适合用大系统分解协调模型求解。模型分两

注：本文为国家"863"计划项目(2002 AA2Z4241)的部分研究内容。

层，第一层(单作物优化)为建立在作物水分生产函数基础上，求解单作物非充分灌溉条件下优化灌溉制度的动态规划模型，其作用是把由第二层(多作物协调)模型分配给第 k 种作物的净灌溉水量 Q_k，在该作物的生育期内进行最优分配；第二层模型为求解水源缺水时多种作物之间水量最优分配的线性规划模型，其作用是利用第一层反馈的效益指标 $F(Q_k)$(最大相对产量)，把有限的总灌溉水量在多种作物之间进行最优分配。模型运行时，首先由第二层分配给第一层每个独立子系统(每种作物)一定水量 Q_k，每个子系统在给定 Q_k 后，各自独立优化，得最优效益 $F(Q_k)$，并将其反馈给第二层。第二层根据反馈的 $F(Q_k)$，计算全系统效益 Z，同时改变上次分配的 Q_k，得到一组新的效益函数 $F(Q_k)$ 及全系统效益 Z，直到求得全系统最优效益 Z_{max} 为止。

在建立数学模型时，作如下假定：整个系统有一个管理机构，可实现水资源的统一管理调配，以实现系统目标最优。对同一种作物，既可单独提取地下水灌溉，亦可调用地表水或与地下水同时灌溉。采用非充分灌溉原理，实行节水灌溉。

2.3.1 单作物优化灌溉制度模型(第一层模型)

2.3.1.1 阶段变量

把全生育期划分为 N 个生育阶段，以生育阶段为阶段变量，$i=1, 2, \cdots, N$。

2.3.1.2 状态变量

状态变量有两个，一是各阶段初可用于分配的水量 q_i(m³/亩)，另一个是各阶段计划湿润层的土壤平均含水率 θ_i，$i=1, 2, \cdots, N$。

2.3.1.3 决策变量

决策变量为各生育阶段的灌水量 d_i(m³/亩)，$i=1, 2, \cdots, N$。

2.3.1.4 系统方程

系统方程有两个，第一个为水量分配方程：

$$q_{i+1}=q_i-d_i+R_i-L_i \tag{1}$$

式中：q_i 为第 i 阶段初可用于分配的水量，m³/亩；d_i 为第 i 阶段的灌水量，m³/亩；R_i、L_i 分别为第 i 阶段可用于分配水量的增加量及其他用水量，m³/亩。

第二个为土壤计划湿润层的水量平衡方程：

$$S_{i+1}=S_i+p_i+d_i-ET_i-C_i-CK_i-K_i \tag{2}$$

式中：S_i 为第 i 阶段计划湿润层土壤平均含水量，m³/亩；p_i 为第 i 阶段有效降雨量，m³/亩；ET_i 为第 i 阶段实际腾发量，m³/亩；C_i 为第 i 阶段排水量，m³/亩，对旱作物其值为 0；CK_i 为第 i 阶段地下水补给量，m³/亩，对地下水埋藏较深地区其值为 0；K_i 为第 i 阶段渗漏量，m³/亩，对采用节水措施进行灌溉时其值可近似假定为 0。

考虑到实测资料的限制，作物耗水量—土壤含水率模型采用线性模型，即实际腾发量 ET_i 与土壤含水率大小成正比。

$$ET_i=ET_{mi}(\theta_i-\theta_{萎})/(\theta_{田}-\theta_{萎}) \tag{3}$$

$$S_i=667\gamma H_i(\theta_i-\theta_{萎}) \tag{4}$$

$$\theta_i=(\theta_{i初}-\theta_{i末})/2 \tag{5}$$

式中：ET_{mi} 为正常灌溉条件下第 i 生育阶段潜在腾发量，m³/亩；$\theta_{田}$、$\theta_{萎}$、$\theta_{i初}$、$\theta_{i末}$、θ_i 分别为田间持水率，凋萎系数，生育阶段初、末和平均土壤含水率(占土壤干重的百分比)；γ 为土壤干容重；H_i 为生育阶段计划湿润层厚度，m。

2.3.1.5 目标函数

作物水分生产函数采用 Jensen 模型，以单位面积实际产量与最高产量比值最大为目标，即：

$$F = \max (Y/Y_m) = \max \prod_{i=1}^{N}(ET_i/ET_{mi})^{\lambda_i} \tag{6}$$

式中：Y 为单位面积实际产量；Y_m 为最高产量；λ_i 为第 i 阶段作物产量敏感性指数。

2.3.1.6 约束条件

$$0 \leqslant d_i \leqslant q_i \tag{7}$$

$$0 \leqslant ET_i \leqslant ET_{mi} \tag{8}$$

$$\theta_{萎} \leqslant \theta_i \leqslant \theta_{田} \tag{9}$$

$$0 \leqslant q_i \leqslant Q_k + \sum_{i=1}^{N}(R_i - L_i - d_i) \tag{10}$$

$$\sum_{i=1}^{N}(-R_i + L_i + d_i) \leqslant Q_k \tag{11}$$

式中：Q_k 为 K 作物全生育期可用于分配的净水量，即协调层分配给第 k 种作物的净水量。

2.3.1.7 初始条件

初始计划湿润层土壤平均含水率为 θ_0，即 $\theta_1 = \theta_0$。

作物全生育期初可用于分配的有效水量为协调层分配给该种作物的净水量，即 $q_1 = Q_k$。

2.3.1.8 递推方程

$$F_i^*(q_i, \theta_i) = \max\{(ET_i/ET_{mi})^{\lambda_i} \cdot F_{i+1}^*(q_{i+1}, \theta_{i+1})\} \quad (i=1,2,\cdots,N-1) \tag{12}$$

$$F_N^*(q_N, \theta_N) = \max\{(ET_N/ET_{mN})^{\lambda_N}\} \tag{13}$$

式中：$F_{i+1}(q_{i+1}, \theta_{i+1})$ 为当前阶段状态为 q 和 θ，决策为 d 时，其后阶段($i+1 \sim N$)的最大总相对产量。

2.3.2 多种作物之间水量最优分配模型(第二层模型)

多种作物之间水量最优分配模型解决地表水(调水)和地下水(浅机井水)在各子系统之间的水量分配问题，同时确定灌区最优种植模式，此模型可用线性规划、非线性规划或动态规划求解，此处采用线性规划模型。

2.3.2.1 变量和时段划分

以年为调节周期，在作物生育期内按月划分时段。本区作物生育期为 1~12 月，共有 12 个时段。按照作物种类划分子系统，其面积分别为 $A_k(K=1,2,\cdots,n)$ 亩，以 $X_{kj}^{(1)}$ 表示 k 作物 j 时段内地表水用量，以 $X_{kj}^{(2)}$ 表示 k 作物 j 时段内地下水用量，单位为 m³ ($j=1,2,\cdots,m=12$)。

2.3.2.2 约束方程

(1)种植面积约束：

$$A_k \leqslant \varepsilon_k A \quad (\forall K) \tag{14}$$

$$\sum_{k=1}^{n} A_k \leqslant A \tag{15}$$

式中：A 为总灌溉面积，亩；A_k 为 K 种作物灌溉面积，亩；ε_k 为 K 种作物灌溉面积占总可灌溉面积的百分比。

(2)地表水约束：各时段地表水引用量不能超过同时段地表水可供用量。

$$\sum_{k=1}^{n} X_{kj}^{(1)} \leqslant SW_j \quad (\forall j) \tag{16}$$

式中：SW_j 为 j 时段地表水可使用量，m³。

(3)地下水约束：一年内含水层总出水量不超过允许开采量。

$$\sum_{j=1}^{m}\sum_{k=1}^{n} X_{kj}^{(2)} - \beta \sum_{j=1}^{m}\sum_{k=1}^{n}[X_{kj}^{(1)} + X_{kj}^{(2)}] + E + R_{out} - R_{in} \leqslant GW \tag{17}$$

式中：β 为灌溉回归系数；E 为含水层年蒸发量，m³；R_{in}、R_{out} 分别为含水层年流出总水量和流入总水量，m³；GW 为综合降水入渗补给后，地下水每年允许开采量，m³。

(4) 需水量约束：各时段作物灌溉水量由各时段地表水和地下水引用量满足。

$$d_{kj}A_k = \nabla_1 X_{kj}^{(1)} + \nabla_2 X_{kj}^{(2)} \quad (\forall\ k,\ j) \quad (18)$$

式中：∇_1、∇_2 分别为地表水和地下水灌溉利用系数。

(5) 非负约束：

$$A_k \geq 0;\quad X_{kj}^{(1)} \geq 0;\quad X_{kj}^{(2)} \geq 0 \quad (\forall\ k,\ j) \quad (19)$$

2.3.2.3 目标函数

以各种作物净效益之和 Z_{max}(元)最大为目标。

$$Z_{max} = \max\left\{\sum_{k=1}^{n} F(Q_k)A_k Y_{mk} P_k\right\} - \sum_{j=1}^{m}\sum_{k=1}^{n}(C_1 X_{kj}^{(1)} + C_2 X_{kj}^{(2)}) \quad (20)$$

式中：$F(Q_k)$ 为第一层反馈的第 k 种作物的效益指标(最大相对产量)；A_k、Y_{mk}、P_k 分别为第 k 种作物的种植面积(亩)、丰产产量(kg/亩)及单价(元/kg)；C_1、C_2 分别为地表水、地下水单位水量使用费，元/m^3。

2.4 模型求解步骤

以每种作物分配的灌溉用水量 Q_k 为协调变量，各子系统满足如下耦合约束。

(1) 初拟协调变量：

$$Q_k A_k = \sum_{j=1}^{m}(\nabla_1 X^{(1)}_{kj} + \nabla_2 X^{(2)}_{kj}) \quad (21)$$

(2) 各子系统最优化。即各种作物各时段的灌水定额 d_{kj} 及各种作物的实际最优效益函数 $F(Q_k)$。

(3) 同步化。因为第一层优化模型以作物生育阶段为时段，第二层优化模型以月为时段，为同步化，需把第一层求得的以作物生育阶段为时段的灌水定额 d_{kj} 转变为第二层以月为时段的灌水定额 d_{kj}'，将各种作物的实际最优效益函数 $F(Q_k)$ 及各种作物各时段的灌水定额 d_{kj}' 分别作为第二层模型目标函数和约束条件的变量值，反馈至第二层。

(4) 第二层平衡协调模型最优化。采用改进单纯形法进行线性规划计算，得各子系统各时段地表水用量 $X_{kj}^{(1)(1)}$ 和地下水用量 $X_{kj}^{(2)(1)}$ 以及各种作物种植面积 $A_k^{(1)}$，同时改变协调变量得新的各种作物的灌溉定额 $Q_k^{(2)}$（各符号的上标(1)表示第一次迭代结果，如 $X_{kj}^{(1)(1)}$、$X_{kj}^{(2)(1)}$、$A_k^{(1)}$ 分别表示地表引水、地下引水和各种作物种植面积的第一次迭代结果）。

(5) 返回第一层再优化。将 $Q_k^{(2)}$ 送入第一层再进行子系统优化，得到新的 E_{kj} 和 $F(Q_k)$，再反馈到第二层进行平衡协调，如此反复迭代，直到结果满足以下精度要求为止。

$$\frac{\|Q_k^{(l+1)} - Q_k^{(l)}\|}{Q_k^{(l)}} \leq \omega_1 \quad (22)$$

$$\frac{\|F(Q_k)^{(l+1)} - F(Q_k)^{(l)}\|}{F(Q_k)^{(l)}} \leq \omega_2 \quad (23)$$

3 应用实例

根据上述模型和基本参数，对示范区多年平均和一般干旱年的地下水、地表水水量年内分配和作物的最优种植面积进行优化，结果见表1、表2。结果表明，现状规划的作物种植面积与优化结果有较大变化，主要是增加了粮食面积，减少了蔬菜面积，这是由水资源量及粮食目标要求确定的。地下水开采量主要集中在汛前(1~5月份占总开采量的89%)，在汛前以开采地下水为主，腾空地下库容，接纳汛期降雨，另外粮食作物小麦、玉米主要在汛前及6月份使用地表水。由于优化了不同时段地表水、地下水的配置，为示范区的用水管理提供了科学依据，所以有效控制了地下水的开采。

表1 羊亭示范区不同作物的优化种植面积

作物	冬小麦	夏玉米	果树	蔬菜	其他
面积(亩)	2 020	2 020	6 048	500	1 512

注：蔬菜为一年两茬。

表2 羊亭示范区不同作物地表水地下水优化分配结果(P=75%)　　（单位：$\times 10^4 m^3$）

项目		月份											合计
		1	2	3	4	5	6	7	8	9	10	11	12	
冬小麦	合计	0	0	0	2.69	2.69	0	0	0	0	0	0	2.69	8.1
	地表水	0	0	0	2.69	2.69	0	0	0	0	0	0	2.69	8.1
	地下水	0	0	0	0	0	0	0	0	0	0	0	0	0
夏玉米	合计	0	0	0	0	0	4.04	0	0	0	0	0	0	4.0
	地表水	0	0	0	0	0	4.04	0	0	0	0	0	0	4.0
	地下水	0	0	0	0	0	0	0	0	0	0	0	0	0
果树	合计	0	0	8.06	0	8.06	0	0	0	8.06	0	8.06	0	32.3
	地表水	0	0	2.88	0	4.87	0	0	0	8.06	0	8.06	0	23.9
	地下水	0	0	5.18	0	3.19	0	0	0	0	0	0	0	8.4
蔬菜	合计	2.67	3.33	3.33	3.67	4.00	5.00	5.00	5.00	5.00	5.00	4.00	4.00	50.0
	地表水	0	0	0	0	0	5.00	5.00	5.00	5.00	5.00	4.00	4.00	33.0
	地下水	2.67	3.33	3.33	3.67	4.00	0	0	0	0	0	0	0	17.0
其他	合计	1.21	1.21	1.21	1.21	1.31	1.31	1.41	1.31	1.31	1.21	1.21	1.21	15.1
	地表水	0	0	0	0	0	1.00	1.00	1.00	1.00	1.00	0	0	5.0
	地下水	1.21	1.21	1.21	1.21	1.31	0.31	0.41	0.31	0.31	0.21	1.21	1.21	10.1

4　结语

利用大系统递阶模型进行区域农业水资源(地表水地下水等)的联合调控，能够解决单种作物灌溉定额和多种作物不同时段获得不同类型水源灌溉水量的共同优化问题，增加了系统应用的灵活性，降低了问题的维数，同时以阶段作物水分生产函数代替全生育期作物水分产量函数，极大地提高了优化结果的精度。

参 考 文 献

[1] 朱道立. 大系统优化理论及应用[M]. 上海：上海交通出版社, 1997.
[2] 郭元裕，等. 灌排工程最优规划与管理[M]. 北京：水利电力出版社, 1994.
[3] 张长江，等. 应用大系统递阶模型优化配置区域农业水资源[J]. 水利学报, 2005(12)：1480-1485.
[4] 李龙昌，等. 多灌溉水源联网调度类型区农业高效用水模式及产业化示范[R]. 山东省水利科学研究院, 2006.

降水和开采对济南市区泉群流量的影响及其贡献

王晓军[1]　陈学群[2]　张维英[3]

(1 济南大学城市发展学院，济南　250002　　2 山东省水利科学研究院，济南　250013
3 山东师范大学附属中学，山东，济南　250014)

就济南泉域的实际情况而言，岩溶水动态的变化往往是受大气降水、开采量、水文地质条件、地下水运移、泉域土地利用变化、城市扩展以及补给区开发活动等多种因素综合影响。本文定量分析大气降水和岩溶水开采对岩溶水动态的综合影响程度以及二者分别对岩溶水动态变化的贡献率。

1 降水的影响

市区泉群的泉流量与同期降雨不一致，存在一定的滞后效应。降水量从 20 世纪 90 年代初期开始增大，泉流量在 20 世纪 90 年代中期也相应出现了多年持续衰减以来少有的回升现象。泉域降水量和泉流量的相关分析表明，1959~2002 年间，泉域范围内当年和前一年的降水对泉流量有较大的影响，且以当年降水为主。此外，根据对济南泉域岩溶水动态资料的进一步分析还可知大气降水对泉域岩溶水动态的影响具有较为明显的分段性(见表 1)：1959~1967 年间泉域的大气降水量相对较多，同一时期，市区年平均泉流量较大，二者动态变化趋势表现出明显的一致性，相关系数为 0.796，接近 $\alpha=0.01$ 的显著水平，此外，这一时期的岩溶地下水位也处于较高水平；1968~2002 年降水量与泉流量进行相关分析，其相关系数皆低于 0.30，均未达到 $\alpha=0.05$ 的显著水平。这说明，1967 年之前降水对泉群流量具有较大的影响。

表 1　不同时期下泉流量与降水量相关关系统计表

时段	当年降水		前一年降水		前二年降水		前三年降水	
	相关系数	显著性水平参数	相关系数	显著性水平参数	相关系数	显著性水平参数	相关系数	显著性水平参数
1959~1967 年	0.796	0.010	0.576	0.105	-0.038	0.922	-0.378	0.304
1968~2002 年	0.290	0.091	0.158	0.365	-0.116	0.507	-0.257	0.136
1959~2002 年	0.340	0.024	0.299	0.049	0.134	0.378	0.130	0.401

2 开采的影响

岩溶地下水开采量和泉流量二者的相关分析也表明岩溶水开采量和泉流量之间存在着明显的负相关：泉流量与岩溶水开采量的相关系数为 -0.862，达到 $\alpha=0.001$ 的显著水平。此外，还表现出明显的阶段性：1959~1967 年间，泉域范围内平均开采岩溶地下水 16.89 万 m^3/d，平均泉流量为 36.66 万 m^3/d，二者的相关系数仅为 -0.320，不足 $\alpha=0.05$ 的显著水平；而 1968~2002 年间，平均岩溶地下水开采量为 66.07 万 m^3/d，平均泉流量仅为 8.22 万 m^3/d，而这一时期二者却呈现出明显的负相关性，其相关系数为 -0.679，达到了 $\alpha=0.001$ 的显著水平。

3 综合影响和贡献率分析

一个要素(因变量) Y 的变化受多种因素(自变量) X_1, X_2, …, X_n 的综合影响，在实际中，不能将所有影响因素定量化，仅能评价出部分影响因素 $X_1, X_2, \cdots, X_n$ 对因变量的综合影响程度，这一综合影响程度可用复相关系数 $R_{y\cdot 12\cdots n}(0 \leq R_{y\cdot 12\cdots n} < 1)$ 来表示：

$$R_{y\cdot 12\cdots n} = \sqrt{1-(1-r_{y1}^2)(1-r_{y21}^2)\cdots(1-r_{yn\cdot 12\cdots(n-1)}^2)} \tag{1}$$

式中：r_{y1} 为 Y 与 X_1 的相关系数，计算公式如下：

$$r_{y1} = \frac{\sum(x_1-\bar{x}_1)(y-\bar{y})}{\sqrt{\sum(x_1-\bar{x}_1)^2 \cdot \sum(y-\bar{y})^2}} \tag{2}$$

$r_{yn\cdot 12\cdots(n-1)}$ 为 X_n 与 Y 之间的偏相关系数。

当 $n=2$ 时：

$$r_{y12} = \frac{r_{y1}-r_{y2}\cdot r_{12}}{\sqrt{(1-r_{y2}^2)(1-r_{12}^2)}} \tag{3}$$

具体到每一个自变量对因变量的影响程度，可用贡献率 $c_i (i=1,2,\cdots,n)$ 来表示：

$$c_i = \frac{r_{yi}}{\sum_{i=1}^{n} r_{yi}} \times 100\% \quad , \quad \text{且} \sum_{i=1}^{n} c_i = R_{y\cdot 12\cdots n} \times 100\% \tag{4}$$

选取 1959～2002 年的降水资料以及同一时期泉域范围内的岩溶地下水开采量资料对泉流量变化作分析。经计算得复相关系数为 0.922，对其进行 F 检验，$F>F_{0.01}$，即 234.18>5.15，故复相关达到了极显著水平。其中，大气降水对泉流量动态变化的贡献率为 26.1%，岩溶地下水开采对泉流量动态变化的贡献率为 66.1%。

由此可见，岩溶地下水开采对泉水喷涌起到关键的作用。

4 结论

对济南市 1959～2002 年的年降水、开采量对泉群流量的影响分析表明，年降水对泉流量的影响很大，此后随着岩溶水开采量的增大，开采量的大小逐渐居主导作用。定量相关分析表明，1959～2002 年间岩溶地下水开采对泉流量动态变化的贡献率远远大于降水的影响，保泉的关键还在于限制泉域范围内的地下水开采量。

参 考 文 献

[1] Ognjen Bonacci. Analysis of the maximum discharge of karst springs[J]. Hydrogeology Journal, 2001 (9): 328-338.

[2] Qian Jiazhong, Zhan Hongbin, Wu Yifeng, Li Fulin, Wang Jiaquan. 2006. Perspective of Fractured-Karst Flow on Spring Protection: A Case Study in Jinan, China[J]. Hydrogeology Journal 14: 1192-1205.

[3] 李砚阁, 杨昌兵, 耿雷华, 等. 北方岩溶大泉流量动态模拟及其管理[J]. 水科学进展, 1998,9(3):275-281.

[4] 陈南祥, 苏万益, 张金炳. 岩溶水系统分析与泉流量预测模型[J]. 华北水利水电学院学报, 1997,18(4):23-27.

[5] 李福林, 马吉刚, 李玉国, 等. 济南市泉群喷涌的控制性参数计算及供水保泉宏观调控措施研究[J]. 中国岩溶, 2002,21(3):188-194.

[6] Li Fulin, Qian Jiazhong, Wang Jiaquan. Temporal change of Jinan Springs discharge and protection of springs. Proceedings of the international symposium on water resources and the urban environment[M]. China Environmental Science Press, 2003. 394-396.

城市湿地生态系统的生态功能与保护对策

王 惠 曲士松 杨宝山 马振民

(济南大学城市发展学院，山东，济南 250002)

1 引言

城市是以人类的技术和社会行为为主导，以生态代谢过程为经络，受自然生命支持系统所供养的"社会—经济—自然复合生态系统"[1]。城市湿地是城市生态系统的重要组成部分，具有城市内其他生态系统不可代替的重要的生态服务功能，对维持城市生态系统的健康和平衡，保障城市的可持续发展具有重要意义。

2 城市湿地的概念与内涵

1971年，全球政府间首次在《湿地公约》中给出湿地的定义，即不论其为自然或人工、长久或暂时性的沼泽地、湿原、泥炭地或水域地带，静止或流动的淡水、半咸水、咸水水体，包括低潮时水深不超过6m的海滩水域。同时，还包括河流、湖泊、水库、稻田以及退潮时水源不超过6m的沿岸带水区[2]。按照此定义，城市内绝大多数的河流和湖泊、池塘等均属于湿地范畴。孙广友等将城市湿地定义为分布于城市(镇)的湿地[3]。

3 城市湿地生态系统的生态功能

3.1 蓄水防洪，净化水质

在城市湿地生态系统中，水的输入来自降水、地表径流、地下水、泛滥河水及潮汐(海滨城市湿地)。水的输出包括蒸散作用、地表外流、注入地下水以及感潮外流。城市湿地的水文特征决定了城市湿地在降水季节分配和年度分配不均匀的情况下，通过天然和人工湿地的调节储存过多的雨水，避免城市发生洪涝灾害，保证城市工业生产和居民生活的稳定水源。另一方面，随着城市工业的发展，工业污染物、有毒物质进入湿地生态系统，在系统中可通过湿地植物、微生物的聚集、沉积作用将水中的营养物质排除，并可富集重金属等一些有毒物质和阻截水中的悬浮物，从而使水体得到改善。湿地较低的pH值，有助于酸水解降解有机物，较严格的厌氧环境也使有机污染物的降解成为可能。

3.2 保护生物多样性

城市湿地的存在为提高城市复合生态系统的生物多样性提供了环境的保障。而生物多样性越丰富，其城市生态系统的稳定性越好。

3.3 调节区域微气候，改善城市环境

观测结果表明，湿地蒸发是水面蒸发的2～3倍，蒸发量越多，导致湿地区域气温越低。湿地的强烈蒸发可导致近地层空气湿度增加，降低周围地区的气温，减少城市热岛效应[4]。

3.4 涵养水源，提供城市居民用水

古往今来城市湿地都是人类发展工农业生产用水和生活用水的主要来源，他在输水、储水和供水等方面发挥着巨大的效益。如济南市的卧虎山及锦绣川两水库每天联合向城市供水10万 m^3，这些水源主要用于生活用水和景观用水。

3.5 美化城市景观，改善居民生活质量

湿地景观是城市景观的重要组成部分。综观世界上的历史文化名城，大都依水而兴。如法国巴黎是因为有了塞纳河才得以兴盛发展，并在河的两岸形成了具有丰富内涵的人文资源。

4 城市湿地生态系统存在的问题

湿地生态系统作为具有独特结构和功能的系统，也是一个不稳定的生态系统。易于受到自然因素或人为活动的干扰，生态平衡易被破坏，并且破坏后的系统很难在短时间内恢复。在近现代城市的发展过程中，由于城市化的加剧，在城市经济快速发展的同时，也导致了对城市湿地生态系统的破坏和不合理的开发利用。

(1)城市化快速发展使城市湿地面积减少，导致了湿地生态系统遭到严重破坏，湿地生态系统涵养水源、调节气候的功能降低，地下水补充能力也受到影响。

(2)由于城市化的加剧，城市人口迅速增长，大量的城市生产和生活废水及有毒物质未经处理直接排入城市湿地，超出了湿地生态系统的自净能力，使城市湿地的水体、土壤受到严重污染。

(3)由于湿地面积减少，湿地景观的破碎化、湿地水质的改变，及人类对外来物种的盲目引入等，改变了城市湿地生态系统的结构，生态平衡遭到破坏，外来物种入侵，物种多样性降低，直接威胁着城市湿地生态系统的生态安全，影响城市湿地生态系统生态功能的发挥。

5 城市湿地生态系统的保护对策

5.1 营建城市湿地公园

湿地公园是指利用自然湿地或人工湿地，运用湿地生态学原理和湿地恢复技术，借鉴自然湿地生态系统的结构、特征、景观和生态过程进行规划设计、建设和管理的绿色空间。湿地公园的建设一方面可有效保护、重建和恢复湿地生态系统，是改善城市生态环境的一个重要内容和有效途径；另一方面，城市湿地公园还是城市景观美学的重要组成部分，它不仅可美化城市形象，还可为城市居民提供一个亲和自然的环境。

5.2 完善监管机制，合理开发利用与自然保护相结合

湿地及其资源管理涉及部门较多，有些河流湖泊等城市湿地的管理甚至涉及多行政区。因此，在城市湿地生态系统管理中要建立和完善科学的湿地监控和功能评价体系，对城市湿地进行长年的测定和调控。

5.3 防止湿地污染

污染主要来自城市工农业生产和城市居民生活中污水和废物的排放。因此，应加强对城市排污场所的管理。具体的对策是在城市规划中，严禁在城市湿地附近开发建设重污染的工业园，对已建立的工业，强行迁出。禁止向城市湿地堆放、倾倒生活垃圾。同时，要进行污水截流，实施雨水、污水分流的城市排水体系，严禁不经处理和未达排放标准的污水直接排入城市湿地。

5.4 加强城市湿地生态系统可持续性的研究

城市湿地生态系统可持续性研究是基于湿地生态学与可持续发展理论结合的交叉研究领域，运用可持续发展的理论来解决城市湿地结构可持续性、功能可持续性等方面的科学问题，构建城市湿地生态系统可持续性评价的科学体系。

5.5 提高全民素质，维护城市湿地环境

城市生态系统是以人为主体的人工生态系统，人既是构成要素，又是最主要的干扰因素。因此，只有让作为城市构成主体的人参与到湿地保护中，才能实现真正的湿地保护。只有加大宣传力度，普及环保教育，提高市民素质，才能更好地保护城市湿地生态系统，发挥城市湿地生态所具有的各种生态功能。

参 考 文 献

[1] 马世骏,王如松. 社会—经济—自然复合生态系统[J].生态学报, 1984, 4(1): 1-9.
[2] 蔡晓明. 生态系统生态学[M]. 北京: 科学出版社, 2002.
[3] 孙广友,王海霞,于少鹏.城市湿地研究进展[J].地理科学进展, 2004, 23(5): 94-100.
[4] 孟宪民. 湿地与全球环境变化[J]. 地理科学, 1999 (5):385-390.

城市化对湿地的影响及保护措施

张明亮

(济南大学城市发展学院，山东，济南 250002)

1 城市湿地的生态和社会服务功能

1.1 城市湿地是城市发展的必要条件

城市湿地不仅为城市居民提供必要的水资源，同时还为城市提供防御、运输、防自然灾害、补充地下水源等服务功能，城市湿地已经成为城市发展的必要条件。

1.2 减缓城市热岛效应，防治城市洪涝灾害

城市热岛效应可增高城市气温。而城市湿地的存在可以明显降低城市热岛效应，湿地水面蒸发和植物的蒸腾作用都很强烈。湿地水平方向的热量和水分交换，使其周围温和湿润。

1.3 城市污染物降解净化作用

随着城市工业的发展，城市环境污染加剧，使城市许多水体趋向富营养化，并在部分湿地中出现了水华。构成水华的主要藻类为蓝藻，而以芦苇、水葱组合的湿地系统和茭白、石菖蒲组合的湿地系统对蓝藻具有一定的去除作用。而且湿地对进入的污染物质具有促进沉积沉降的自然特性。污染物质在湖体中随沉积物沉降后，通过湖中的湿生植物吸收，经化学和生物化学转换而被储存起来，从而达到污染净化的作用。

1.4 为动植物提供丰富多样的栖息地，保护生物多样性

湿地由于生态环境独特，决定了其生物多样性的特点。湿地鸟类种类、鸟类总数、物种多样性都高于周围地区。

1.5 为城市居民提供休闲娱乐场所和教育场所

湿地丰富的水体空间，水面多样的浮水和挺水植物，以及鸟类和鱼类，都充满大自然的灵韵，河流两岸、河心沙洲的自然景观为人们提供了亲近自然的场所。另外，湿地丰富的景观要素、物种多样性，为环保宣传和对公众进行相关教育提供了场所。

2 城市化对城市湿地的影响

2.1 湿地面积大幅度减少

在城市化进程中，城市水面率逐步降低，不透水地面积逐步扩大，导致城市湿地系统逐步消失。城市湿地的萎缩使城市地区的热岛效应、降雨强度明显加大，地表水注蓄和下渗能力减弱，对城市防洪造成了严重的威胁。

2.2 湿地破碎度加大，连通性降低

城市建设用地的扩张使湿地景观破碎度加大。城市湿地本身面积较小，破碎度加大，连通性降低，使得城市湿地呈现孤岛状，不利于物种地交流和湿生态服务功能的发挥。

2.3 湿地污染严重

由于城市化过程中的不合理规划，加上大量的生活和工业用水的排放，湿地水质下降，受污染情况愈来愈严重，并对其周围环境也造成污染，降低了湿地的生态及社会服务功能。

2.4 生物入侵破坏湿地生境

因湿地生境条件的改变以及湿地规划中盲目引进一些异地物种，引起外来物种对湿地生境的

入侵，降低了本地物种的存活几率，增加了城市湿地保护和恢复的难度。

3 城市湿地的保护措施和规划途径

从 20 世纪的 70 年代起，欧美、日本一些国家开始重视对城市河流湿地的保护，他们将部分已经被破坏的城市河流湿地逐渐地进行回归自然的修复，在充分利用自然地形、地貌的基础上，建立起阳光、植物、生物、土壤、堤体之间和谐共存的城市河流生态系统。

3.1 保持湿地景观的自然性，推动自然型河流建设

首先要维护和恢复河道和海岸的自然形态，确保湿地生态服务功能的有效发挥。但目前存在的弊端具体表现为水泥护堤、河床破坏、裁弯取直、拦河筑坝等。

水泥护堤使得水的自净能力消失殆尽，水—土—植物—生物之间形成的物质和能量循环系统被彻底破坏；河床破坏后切断了地下水的补充通道，导致地下水位不断下降；自然状态下的河床起伏多变，基质或泥或沙或石，水流或缓或急，为多种水生植物和生物提供了适宜的环境，而水泥衬底后的河床，异质性不复存在，许多生物无处安身；河流裁弯取直降低了河流蓄洪涵水和削弱洪水的能力，破坏了河流生境的多样性，也破坏了河流独有的自然形态美和为人类提供富有诗情画意的感知体验空间。

3.2 建立水系廊道网络，增加湿地的连通性

各种孤岛式的湿地斑块之间的连通性急剧下降，破坏了景观格局的连续性，严重阻碍了湿地生态效应的发挥。因此，在城市湿地生态恢复和重建时，应在湿地斑块之间增加水系廊道，使各个孤立湿地斑块形成网状结构。

3.3 实施污染控制工程

实施科学的方法杜绝和减少污染源是进行城市湿地保护的必要前提。对此，一方面，要迁出城市湿地附近的污染工业，禁止向湿地堆放、倾倒生活垃圾，从根本上消除污染源；另一方面，要进行污水截流，实施雨水污水分流的城市排水体系，严禁不经处理和未达到排放标准的污水直接排入城市湿地。

3.4 建立及恢复城市湿地时应以本地乡土物种为主

乡土物种是通过多年的物种选择证明适宜生长于本地生境的物种。在恢复及建立湿地植被时，利用本地物种不仅加大了实际工作的可行性，也可节约自然资源及社会资源。

3.5 恢复重建湿地，确保湿地面积

通过生态技术或生态工程对退化或消失的湿地进行修复或重建，再现干扰前湿地生态系统的结构和功能以及相关的物理、化学和生物学特性，使其发挥应有的作用，对破坏严重的湿地，通过园林绿化工程和植物群落重建，可加快湿地植被的恢复。同时，为保护本地生物多样性，城市湿地的恢复和重建应注意确保城市湿地斑块面积。

3.6 建立持续的城市湿地监控机制

在湿地生境退化和丧失较为严重的区域，要完全恢复功能健全的湿地一般需要经过 10~15 年，而且湿地系统各项功能的发育速度有所不同。城市湿地的保护和恢复还必须建立相应的监控机制和功能评价体系，以对城市湿地进行持续的测定和调控。

参 考 文 献

[1] 李文华, 欧阳志云, 等. 生态系统服务功能研究[M]. 北京: 气象出版社, 2002.
[2] 俞孔坚, 潮洛蒙, 等. 城市生态基础设施建设的十大景观战略[J]. 规划师, 2001 (6): 9-13.
[3] 潮洛蒙, 李小凌, 等. 城市湿地的生态功能[J]. 城市问题, 2003 (3): 9-12.

某饮用水源水库铁、锰垂直分布规律及原因探讨

王海霞

(济南大学城市发展学院，山东，济南 250002)

水源存在季节性的水质变化现象，如 Fe、Mn 离子在一定时间内浓度大幅度升高，能够达到平均浓度的 10 倍左右[1]。对水库铁、锰垂直分布规律进行探讨，探明其内部反应具体过程机理，对于采取针对性措施提高水源管理水平和改进供水水质，为水厂寻求获得较好水质的取水方式提供依据。

本文研究了揭阳市某水库中温度、溶解氧、铁、锰、硫酸盐和总磷等水质参数的分布规律，探讨了水体中铁、锰的分布规律。

1 水库水温结构

水温结构判别，采用库水替换次数的 α 指标：

$$\alpha = \frac{\text{多年平均径流量}}{\text{水库总库容}}$$

当 $\alpha<10$ 时，水库水温为稳定分层型；当 $10<\alpha<20$ 时，水库水温为过渡型；当 $\alpha>20$ 时，水库水温为混合型。

$$\beta = \frac{\text{一次洪水总量}}{\text{水库总库容}}$$

当 $\beta<0.5$ 时，洪水对水温结构无影响；当 $0.5<\beta<1.0$ 时，呈过渡阶段；当 $\beta>1.0$ 时，洪水对水温结构有影响。

该水库多年平均流量 1 255.5 万 m^3，总库容 1 260 万 m^3，$\alpha=0.99$，所以水库水温结构为稳定分层型。水库一次洪水总量为 834.8 万 m^3(3 d)，$\beta=0.66$，因此洪水对水库水温结构影响不大。

2 试验方法

试验采样点分别设在水库的进水口、中心区和出水口。于 2002 年在水温分层的夏秋季及水温不分层的冬春季进行取样分析。

pH 值、DO 值的测定采用常规分析方法，Fe、Mn 值的测定采用原子吸收法，Fe、Mn 的检出线分别为 0.03 mg/L 和 0.01 mg/L，标准偏差分别为 0.86%和 0.85%。

3 试验结果及讨论

3.1 水温分层与 pH 溶解氧的变化关系

9 月份气温较高，水体呈现稳定分层状态，上层水温为 27.4℃，由于大气供氧及藻类的光合作用吸收 CO_2 放出氧气，因此 DO 值较高为 7.5 mg/L；温跃层水温为 24.7 ℃、DO 值为 3.68 mg/L，下层水温为 22.4 ℃，DO 值降到 0.6 mg/L。上下层水体被温跃层分开而缺少对流运动，导致溶解氧无法穿过温跃层，使中下层水体缺氧呈现还原状态。秋末冬初表层水温随气温逐渐下降，上层温度为 24.6℃，中层为 24.5℃，下层为 21.9℃，水库中可以形成对流，从而为中下层水体增加了

溶解氧，使中下层水体逐渐转入氧化状态。1月份表层水温继续下降，上层温度为18℃，中层为16.9℃，下层为15.6℃，库水对流运动达到最强，水体处于氧化状态。因此，水库的中下层水体一年四季重复呈现出氧化还原状态。

3.2 水库铁、锰垂直分布特征

9月份表层水铁、锰含量比较低，Fe离子浓度为0.017 mg/L，Mn离子浓度为0.02 mg/L；在中水层中铁、锰浓度发生明显上升，Fe离子浓度为0.124 mg/L，Mn离子浓度为0.207 mg/L；在下层处属于厌氧区，铁锰浓度都达到了最高值，Fe离子浓度为1.112 mg/L，Mn离子浓度为0.732 mg/L。深冬表层水铁、锰含量仍然比较低，Fe离子浓度为0.015 mg/L，Mn离子浓度为0.005 mg/L；在中水层水体中铁、锰浓度继续保持高值，Fe离子浓度为0.12 mg/L，Mn离子浓度为0.123 mg/L；在下层处，属于厌氧区，铁锰浓度都达到了最高值，Fe离子浓度为1.105 mg/L，Mn离子浓度为0.438 mg/L。由此可见，无论是夏末还是初秋、深冬，深层水体铁、锰的含量都保持高值，平均浓度分别是表层水体的69.2倍、33.4倍。

4 原因分析

4.1 基岩对水库铁、锰在垂向分布的影响

水库水源供给部分来自地下水，因此水库中铁、锰物质来源主要是地下水对基岩的溶解，溶解性铁、锰的形成与地下水的补给、径流条件和水库地球化学环境等因素有关。建库后，库底的地球化学环境发生了变化，库底水中的CO_2含量增加，库底沉积的有机物质分解产生有机酸，促使水库底层水的pH值、Eh值降低，库底成为还原环境或弱还原环境。地下水总体上是呈饱和水缓径流状态，有利于基岩岩石中某些化学成分的溶解。还原环境使环境含有像铁、锰这些变价元素的化合物的溶解度增大。在地下水中普遍溶解有铁、锰等成分，它们在地下水中呈胶体的形式存在[2]。由于水库的表层环境与深部不同，温度比也深层高，同时表层水中的含氧量也比中下层的含氧量高，原来在深部是还原环境而到了表层就为氧化环境，于是在水中的低价铁、锰到表层会被氧化，形成高价的铁、锰而产生沉淀。

底部铁、锰浓度较大主要是底部水处于还原环境，有利于低价铁、锰的溶解，引起底层水对基岩产生了化学潜蚀的结果。

4.2 季节性缺氧对铁、锰垂向分布的影响

秋冬季水温在垂向上大致趋于等温状态，对流运动比较强烈，整个水体处于氧化状态。此时铁主要以三价铁、锰以四价锰的形态存在，分别为$Fe(OH)_3$和MnO_2都不溶于水，迁移能力很低，逐步沉积于库水底部，在沉积物—水界面附近沉淀[3]，并储积于沉积物表层，导致沉积物铁、锰含量的增高。水中的铁、锰的氧化还原反应随季节性溶解氧的变化而变化，秋冬季水体溶解氧比较高，水体偏碱性，水库水动力学稳定，铁、锰的氧化还原反应循环在底部完成，使铁、锰在底部富集，而不会向上水体释放。

秋末，当水体形成温跃层时底部溶解氧不易得到补充；当水库受到污染时，有机物和氨氮以及其他耗氧性物质浓度升高，加剧溶解氧的消耗，容易形成底部厌氧环境。在厌氧条件下，使不溶性的$Fe(OH)_3$和MnO_2转化为溶解性Fe^{2+}和Mn^{2+}，由于浓度差而向表层水体扩散，形成沿水体深度方向从表层向深层浓度逐渐升高的现象。温跃层以下的水体逐渐因缺氧而呈现还原状态，水体偏酸，沉积物中的铁、锰的氧化还原反应在沉积物与界面水之间进行，这时沉积物中的高价铁、锰被有机物还原为低价而溶于上覆水中，其迁移能力很强：

$$Fe^{3+} \rightarrow Fe^{2+}$$

$$MnO_2 + 4H^+ \rightarrow Mn^{2+} + 2H_2O$$

当溶解性的Fe^{2+}和Mn^{2+}经过扩散，进入溶解氧浓度比较高的表层水体时，能够被氧分子氧化，重新形成不溶性的$Fe(OH)_3$和MnO_2，部分又以颗粒的形式，沉淀返回深层水体的厌氧环境。而

返回来的铁和锰,可能又被重新还原成为溶解性的 Fe^{2+} 和 Mn^{2+},通过扩散重新进入上部好氧层,然后又被氧化和沉淀,形成了一个完整的循环,这个循环过程通常称为"车轮反应"[4]。释放过程从夏季开始,与沉积物接触的界面及下层水铁、锰浓度逐渐增高,到秋季水体翻转化以前铁、锰释放扩散达到高峰,中下层铁、锰浓度升高,但水面至温跃层顶部的水体由于风力搅动、大气复氧及藻类光合作用增氧,始终处于氧化状态,故铁、锰浓度仍比较低。因为夏秋季铁、锰向界面水扩散高于冬春季,夏秋季铁、锰浓度要高于冬春季的铁、锰浓度。

4.3 沉积物—水界面对铁、锰循环的控制作用

水库中铁、锰围绕着沉积物—水界面形成的循环,由还原—扩散—氧化—沉积四个环节组成[5]。沉积物中的铁、锰氧化物充当有机质降解的氧化剂被还原溶解,溶解态铁、锰通过孔隙水向上覆水体扩散迁移,在沉积物表面重新被氧化成铁、锰氧化物沉淀在界面上,形成微粒态铁、锰氧化物富集。但是水库中的铁、锰循环不仅仅受沉积物—水界面制约,因为在非缺氧季节里,氧化还原边界层位于沉积物—水界面附近,铁、锰界面循环的结果是沉积物表层铁、锰的富集;当水库季节性缺氧时,氧化还原边界层由沉积物向上覆水体季节性迁移,铁、锰界面循环也向上覆水体扩展,结果是沉积物表层铁、锰的季节性释放和库水缺氧后铁、锰的富集[6],因此在缺氧季节的铁、锰浓度高于非缺氧季节的铁、锰浓度。

5 结语

(1)水库铁、锰垂直分布随季节有一定的变化,但不同于其他水库或湖泊在非缺氧季节,铁、锰的分布在垂向上基本一致。而该水库中的铁、锰浓度即使在非缺氧季节也是从表层向深层浓度逐渐升高的现象,出现这一现象的原因主要受水库基岩的影响。

(2)根据水库铁、锰垂直分布规律,上层铁、锰浓度较低,达到《地表水环境质量标准》,下层水严重超标。因此,可将水厂取水口设在上层取水,从水源环节来改善自来水厂的处理效率。

参 考 文 献

[1] 邓晓林,王国华,任鹤云.上海城市污水处理厂的污泥处置途径探讨[J].中国给水排水,2000,16(5):19-222.

[2] Taillefert M, Macgregor B J, Gaillard J F. Evidence for a dynamic cycle between Mn and Co in the water recolumn of a stratifiled lake [J].Environ Sci Technol, 2002(36): 468-476.

[3] 施希京,彭汉兴.水电站坝基析出物的形成研究[J].水文地质工程地质,1995(3):43-46.

[4] 袁文权,张锡辉,张光明,等.低浊水源铁锰与磷耦合转化研究[J].给水排水,2003,29(11):10-14.

[5] 罗莎莎,万国江.云贵高原湖泊沉积物—水界面铁、锰、硫体系的研究进展[J].地质地球化学,1999,27(3):47-52.

[6] 万国江.环境质量的地球化学原理[M].北京:中国环境科学出版社,1988.